# ENVIRONMENTAL POLITICS
*FROM SOCIABILITY TO SUSTENANCE*

# ENVIRONMENTAL POLITICS

## *FROM SOCIABILITY TO SUSTENANCE*

***By***

**Dr. Prasenjit Maiti**

*Assistant Professor*
*Public Systems Management and*
*Human Resource Management*
*Department of Indian Institute of*
*Social Welfare and Business Management*
*Management House*
*College Square*
*West, Kolkata – 700 073*

**DPH**

**DISCOVERY PUBLISHING HOUSE PVT. LTD.**
**NEW DELHI-110 002**

First Published-2010

ISBN 978-81-8356-653-7



*Published by:*

DISCOVERY PUBLISHING HOUSE PVT. LTD.
4831/24, Ansari Road, Prahlad Street
Darya Ganj, New Delhi-110002 (India)
Phone: +91-11-23279245, 43764432 • Fax: +91-11-23253475
*E-mail:* parul.wasan@gmail.com
info@discoverypublishinggroup.com
*web:* www.discoverypublishinggroup.com

*Printed at:*
Sachin Printers
Delhi

# Preface

Environmental politics is an important issue at both the policy and praxes levels, as it leads to critical concerns like political decentralization, good governance, political ecology, legal reforms, institutional re-explorations, fiscal readjustments, eminent domain and the overall ethics of what we tend to address as environmentalism as such. This issue has recently emerged in the context of the United Nations Development Program's Millennium Development Goals as a *problematique* of Globalization/Glocalization that may be ideally negotiated in terms of locally workable strategies. Environmental management indicators can be both quantitative and qualitative in nature in order to provide primary inputs to monitor different projects, evaluate progress of project implementation and guide the stakeholders to sustain the benefits of these projects. Such indicators may also evaluate the impacts of these projects on the quality of life of the stakeholders before and after implementation, and also examine the increased level of community participation and awareness-generation *vis-à-vis* project interventions. Participatory environmental development may be defined as an interactive process during the passage of which the concerned stakeholders gain access to effective decision-making either directly or via institutionalized dynamics. This helps to identify the needs and perceptions of the primary stakeholders, sustains project benefits even after withdrawal of the implementing agencies, monitors physical progress, streamlines civil societal institutional arrangements and develops a sense of ownership among rural community members vis-à-vis their collective assets. Environmental management is generally an arrangement of value-specific interventions where progress indicates contrasting worldviews, where new actors help evolve different modes of production and where dynamic stakeholder interfaces result in participatory mechanisms. Environmental politics so tends to emerge as an ongoing process that tries to resolve (with uneven success) the tension of priorities between the center and periphery. This Volume would be of interest to social scientists, development consultants, government officials, environmental scientists/managers, policy-makers and planners.

**–Author**

# Contents

# List of Contributors

1. **Prasenjit Maiti**, *prasenjit.maiti@gmail.com* is a political sociologist. He taught at the Department of Political Science of the University of Burdwan during 1995-2002 before joining the Development Sector in 2003. He was on secondment as a Doctoral Fellow at the Institute of Federalism under the University of Fribourg, Switzerland in 1999. Dr Maiti also worked for the Forum of Federations, a Non-Government Organization based at Ottawa, Canada in 2003. Dr Maiti later served as an Evaluation Consultant to TERI while working on the National Green Corps Project funded by the Ministry of Environment and Forests, Government of India. He is also Referee for Environment & Development Economics (University of York, Heslington) published by the Cambridge University Press. His first book focused on the theories and praxes of governance in the Indian context while his second and third books dealt with development discourses and ethical applications with special reference to developing societies. His present research interests include civil society, social capital, political ecology, participatory development, urban poverty, Structural Adjustment Programs, politics of aid and choices of development. His latest book (forthcoming) would stress upon the problems of development ethics in the context of good governance and involuntary

displacement induced – among other related factors - by the dynamics of infrastructure development. Dr Maiti has of late tended to specialize in the critique of sustainable development as a rhetoric of the top-down Eurocentric / Anglophonic approach. He was invited as a Guest Professor to the Danish Institute of Rural Research and Development (IFUL) under the University of Southern Denmark at Esbjerg in 2007. He presently teaches at the Indian Institute of Social Welfare and Business Management (IISWBM) Kolkata. Dr Maiti was recently invited to the Sweden Conference on Urban Policies and Social Capital 2009 at Lidköping as a Foreign Expert.

2. **Aminu Mamman,** *aminu.mamman@manchester.ac.uk* is Management and Organizational theorist. Dr Mamman is currently based at the Institute for Development Policy and Management of the University of Manchester, UK. He had earlier taught at Monash University, Australia.

3. **Christopher Rees,** chris.rees@manchester.ac.uk> is a Chartered Psychologist. He held HR positions in several public and private sector organizations before joining the Development Sector. Dr Rees is now based at the Institute for Development Policy and Management of the University of Manchester, UK. He has published widely on issues relating to human resources and organizational change.

4. **Arunayan Sharma,** *arunayan@gmail.com* is a qualified Civil Engineer with a background in Environment Management. He also has a post-graduate degree in Environmental and Ecological Sciences and an LLM in Environmental Law. He is an amateur ornithologist and environment /wildlife photographer. He is also a freelance science, environment and wildlife writer of *The Statesman*.

5. **Bibi Khadiza,** khadiza_urp@yahoo.com is an Urban Planner. She studied in the Urban and Rural Planning Discipline of the Science Engineering and Technology School of Khulna University, Bangladesh. Her academic

interests include development issues, social research and technological adaptation in rural cultures. Her Undergraduate Thesis focused on the problems of landuse conflict in rural livelihood pattern.

6. **Mohammed Shariful Islam,** *sharifurp02@gmail.com* is a Researcher on Urban and Regional Development. He studied in the Urban and Rural Planning Discipline of the Science Engineering and Technology School of Khulna University, Bangladesh. His present interests include development issues, social research and geoinformatics. His Undergraduate Thesis focused on the problems of accessibility to opportunities due to work activity pattern of individuals.

7. **Khan Rubayet Rahaman,** *krrporag@yahoo.com* is an Urban Planner and a Faculty Member of the Urban and Rural Planning Discipline at Khulna University in Bangladesh. Dr Rahaman obtained his Masters Degree from KTC, Sweden and his Doctoral Degree with a JICA Fellowship from the University of Tokyo, Japan in 2006. His academic interests cover a wide range of social scientific issues.

8. **Chandima Daskon,** *dasch015@student.otago.ac.nz* is is a Development Geographer. She holds a Bachelor of Arts (Honours) Degree in Geography from the University of Peradeniya, Sri Lanka and a Masters Degree in Social Change and Geography from the Norwegian University of Science and Technology (NTNU). Currently she is reading for her PhD at the Department of Geography of the University of Otago, New Zealand. Her research focuses on the 'Sustainable Livelihoods and Rural Development' with special emphasis on the role of 'rural culture' as a vital resource for attaining rural sustainability. Before embarking on her PhD study, Daskon was involved in a research project with the United Nations University (UNU), Japan and also worked as a University Lecturer in Geography in Kalaniya University and Sri Jayewardenepura University in Sri Lanka.

9. **Deane Neubauer,** *deanen@hawaii.edu* is Professor Emeritus in the Department of Political Science at the University of Hawaii, Manoa, Senior Research Fellow of the Globalization Research Center (UHM) and Senior Advisor to the Education 2020 Program of the East West Center. Educated at the University of California, Riverside and Yale University, he has taught additionally at the University of California (Berkeley and Irvine), Waikato University (NZ) and the University of Sydney. His research interests lie in health policy, political economy and globalization. His work explores globalization phenomena as a major vector of social change throughout the world. He served as the founding dean of the College of Social Sciences at the University of Hawaii, Manoa (1980-88,. In 1999 he founded the Globalization Research Center at the University of Hawaii, Manoa and subsequently the Globalization Research Network, a collaboration of four US universities. He served as Interim Chancellor of the University of Hawaii, Manoa from 2001-2002 and as Interim Vice- President for Academic Affairs for the University of Hawaii System from 2001-2004.

10. **Dimitar Eftimoski,** *dimitar@ukim.edu.mk* earned his PhD in 2002 in the field of development economics with a thesis on "Human capital, economic development and human development". He lectures at the Faculty of Economics at the European University in Skopje, Faculty of Business Economics in Skopje and at the Postgraduate Studies Level of the Institute of Economics at the Ss. Cyril and Methodius University in Skopje. He is the author of an introductory textbook, *Economics of Development* (Institute of Economics - Skopje). His research includes work on human development, economic development, economic growth, poverty and inequality.

11. **David L Debertin,** *ddeberti@uky.edu* is Professor of Agricultural Economics at the University of Kentucky in Lexington, Kentucky where he has been employed since 1974. He completed his PHD in 1973 at Purdue University.

Dr Debertin has conducted research and taught courses dealing with rural economic development and the role of social capital in affecting the course of rural development. He has written over 50 refereed journal articles dealing with economic development, the economics of education and rural social capital. He holds teaching awards from both the Southern Agricultural Economics Association as well as the University of Kentucky Excellence in Undergraduate Instruction Award.

12. **Stephan J Goetz,** *sgoetz@uky.edu* is Director, The Northeastern Regional Center for Rural Development and Professor of Agricultural and Regional Economics at the Pennsylvania State University, University Park, PA 16802-5602. He was formerly professor of agricultural economics at the University of Kentucky. Dr Goetz has published or presented over 150 professional papers on rural economic development topics, including 10 book chapters, and he is the senior co-editor of two books. He is the principal investigator on external grants valued at over $2.5 million. Among many other service roles, he is past chair of the Social Sciences Sub-Committee of the Experiment Station Committee on Organization and Policy (ESCOP), and he is a member of the Board of Directors of the Center for Rural Pennsylvania, a legislative agency of the PA General Assembly.

13. **Louise Warwick-Booth,** *l.warwick booth @ leedsmet.ac.uk* is a Lecturer in Health Policy at Leeds Metropolitan University, UK. She was previously a Graduate Tutor in Sociology, Social Policy and Criminology at the University of Sheffield, where she completed a PhD in community-based research and evaluation within social regeneration. Dr Warwick-Booth's current research interests include community development, community research and social inclusion.

14. **V Anil Kumar,** *anilkumar@isec.ac.in* is presently Assistant Professor at the Centre for Decentralization and Development of the Institute for Social and Economic

Change (ISEC), Bangalore, India. He obtained his M.Phil and PhD degrees in Political Science from the University of Delhi and MA in Political Science from Jawaharlal Nehru University. His areas of interest include, Democratic Decentralization and Rural Institutions; Agrarian Change and Rural Development; Public Policy and Policy Processes; Social and Political Theory. He has considerable experience in the Development Sector as well as in academia.

15. **Ranabir Dey,** *ranabir_dey@rediffmail.com* did his Master of Philosophy in Planning and Development as a Research Scholar from the Department of Humanities and Social Sciences of the Indian Institute of Technology Bombay. He is interested in the area of urban issues related to governance and sustainable urban management and planning. He has written technical papers for presentation on 'Patterns of urban transport development in Kolkata' (March 2006), 'Role of healthy cities' (March 2007) and 'Community safety plan' (November 2006) at National-level Conferences. He was employed as an Urban Planner in the DFID funded Kolkata Urban Services for the Poor Program in Kolkata where he was assigned the role of formulating the Draft Development Plan of an Urban Local Body.

16. **Kaushalendra Prakash Goswami,** *kp_goswami 2004@yahoo.co.in* joined the Department of Geography of Benaras Hindu University, India as a Lecturer in 2004. His Doctoral Thesis was on the 'Impact of ecology and environment on forests and agriculture: A case study of Gaya District'. Dr Goswami has specialised in Environmental Geography and Resource Planning.

17. **Julia Maturana**, *jmaturana@usat.edu.pe* is an environmental economist. She worked at the Centre for International Forestry Research (CIFOR) headquarters in Indonesia, where she carried out this research. Professor Maturana is now a associated with the Research Centre for Sustainable Development (CIDES) in Peru,

where she also teaches economics at USAT University. Previous publications focused on smallholder forest Plantations in the tropics, plantation forestry in Java, Economic Costs and Benefits of Allocating Forest Land for Industrial Tree Plantation and Diversity and Importance of Natural Resources for local communities. Her present research interests include poverty, social development, and environmental policy.

18. **Cristophe Simon**, *c_simon70@yahoo.fr* is a graduate student from the French Institute of Forestry, Agricultural and Environmental Engineering (ENGREF) of Montpellier. He worked in Indonesia on a short-term research project to collaborate with Professor Julia Maturana under her supervision.

19. **Irmayanti,** *Helda76@Yahoo.com* is a graduate student from the Bogor Agricultural University (Institut Pertanian Bogor) of Indonesia. She worked for CIFOR on a short-term research project to collaborate with Professor Julia Maturana under her supervision.

# 1

# An Introduction to Environmental Politics

*Prasenjit Maiti*

The present volume lays emphasis on several theoretical and empirical aspects of environmental politics so that the critical dependencies between human and physical environmental settings are suitably focused and discoursed in the process. The social issues and concerns of environment have become more prominent and relevant especially since the World War II, when the impact of weapons of mass destruction and nuclear holocaust made us painfully aware about the politics and economics and sociology of environment. The alternative to sustainable environmental development at the beginning of the new millennium appears to be mutually assured destruction. So new disciplines such as human and political ecology have developed in order to interrogate the politics of development that is more often than not an interplay between the choices of policymakers and voices of stakeholders caught desperately between the currents of so-called human progress.

Sustainable environmentalism, moral ecosophy and political ecology are critical concerns of the new millennium at the policy and praxes levels, as this issue predicates new and exciting disciplines such as political ecology, environmental ethics, human ecology, eco-feminism, environmental law, moral ecosophy and development ethics. Sustainable environmental development has of late emerged as a serious *problematique* that requires to be negotiated in terms of workable strategies.

Environmental Management Plans generally constitute the formulation of quantifiable benchmarks supported by Monitoring and Evaluation exercises to assess the benefits accruing from implementation of EMPs. Monitoring is the scientific collation and compilation of primary as well as secondary data to gain insights into the process, project and progress indicators while Evaluation is the systematic analysis of environmental projects, including aspects like design, implementation and results. Evaluation leads to increasingly informed decisions, allowing those involved in this process to learn from experience and to be accountable to the dynamics of the project. Monitoring and Evaluation should be utilized to establish accountability and understand the dynamics of environmental projects.

Information, Education and Communication strategies involve different concerns of environmentalism related to awareness-generation, advocacy/activism and sensitization that are attuned to culture and gender specifics. In order to scientifically implement project-related tasks it is rather imperative to develop a time-frame of IEC activities to be carried out during environmental project cycles. Any given Environmental Action Plan must elaborate and schedule the tasks to be undertaken in accordance with the aims and objectives of IEC as mandated by the scope of different environmental projects.

Benefit Monitoring and Evaluation of environmental projects in terms of new physical resources created is an exercise related to measurement of technical interventions and designs at the levels of planning and implementation. It is important to understand that Impact Assessment of environmental projects should also include the ethos and best practices of participatory development.

Participatory environmentalism can be sustained by a basket of methodological tools such as Focus Group Discussions, Gender Interviews, Community Meetings, Logical Framework Analyses and Risks Assessment. Resource Mapping, for instance, is an imperative requirement for Joint Forest Management and Natural Resource Management initiatives. Needs assessment of physical resources as perceived by the community must be based on such participatory methods. This is especially true in the case of rainwater harvesting and arsenic-free drinking water projects. Micro-planning at the grassroots should inform the choices and voices of the concerned stakeholders. It remains a fact that sustainable environmental management is nearly always intrinsic to local-level realities. So it is critical to be attuned to the local for environmental movements to succeed at the global level.

Environmental management indicators can be both quantitative and qualitative in nature in order to provide primary inputs to monitor different projects, evaluate progress of project implementation and guide the stakeholders to sustain the benefits of the projects. Such indicators may also evaluate the impacts of the projects on the quality of life of the stakeholders before and after implementation and also examine the increased level of community participation and awareness-generation vis-à-vis project interventions. Under different Initial Environmental Examination scenarios the State of Environment may be studied in the context of varying Pressure State Response Models.

Environmental Action Plans must also lay down strategies for both qualitative and quantitative methods in order to freeze the ground realities and design physical interventions under different environmental projects. Such interventions should generally be participatory, i.e. underpinned by the Stakeholders' Consultative Process. SCP is central to the methodology of environmental projects. SCP moves away from the conventional top-down approaches that involve measurement of project performance against pre-set indicators by using standardized procedures and tools. The new strategy of participatory development lays its primary emphasis on Public Information and Consultation endeavours.

Environmental projects underpinned by the rationale of participatory development should define statements of measurable objectives in terms of physical interventions, design a structured set of indicators covering outputs and likely environmental impacts and include provisions for data analyses and periodic reporting to the project stakeholders. Process indicators are benchmarks that are concerned with the dynamics of environmental change, i.e. how change is to be effected at the grassroots, how the process of change is to be managed with the help of external interventions and internal facilitation, how best change can be reconciled with conventions?

Environmental projects cannot always follow predetermined approaches. Best practices of environmentalism more often than not emerge from chaos but subsequently lead to cohesion by resolving the tension between priorities and utilities. BME indicators may be designed keeping in view the fact that environmental projects aim to facilitate need-based awareness-generation programmes. Participatory environmentalism can be defined as an interactive process during the passage of which the concerned stakeholders gain access to the decision-making process either directly or via institutionalized representation. This helps to identify the needs and perceptions of primary stakeholders, sustains civic amenities after withdrawal of implementing agencies, monitors progress of physical works, streamlines community-level institutional mechanisms and develops ownership among community members vis-à-vis collective assets and Common Property Resources.

Environmental participatory strategies operate at two levels: [1] the communitarian level where the community is brought back into environmental focus in order to build and sustain a proprietary sense among primary stakeholders and [2] the neoinstitutional level where "actors and institutions" are transformed into "actors in institutions" i.e. end-users of environmental resources are mobilized suitably to identify their community agenda with environmental interests. These are strategic measures utilitarian tools to motivate the community to become proactive and safeguard collective concerns at the level of the environment. Environmental awareness programs serve to build empathy at the community-level and facilitate dialogic rather than pedagogic terms of discourse at the project implementation, monitoring and evaluation levels. Community mobilization and participation activities are not an end but a means to secure rapport-building that may ensure acceptance of environmental initiatives. Environmental interventions must be scientifically implemented and informed by a definitive plan of action. This is despite the fact that environmental monitoring and management generally emerge from flux and can seldom be planned in a structured manner.

Environmental problems may be identified and cause-effect relationships determined in this process. This may also help to identify benchmarks of solution impacts (facilities proposed to be provided under environmental projects). The problems identified should be converted into need-based statements. This exercise requires discussion and prioritization of environmental needs. The achievements of environmental projects generally depend on how project beneficiaries react to the resources created or sustained by the project initiatives. For participatory approaches to MandE to be effective, these need to be integrated with environmental projects and adopted on a continuous basis. These results can be utilized restrategize the key issues and generate new questions during subsequent MandE. Environmental projects may simultaneously develop new activities and encounter new problems that which may affect the stakeholders' perceptions of project-related activities. Communities may also experience changes from other sources that may have an impact on the progress of environmental projects.

Environmental Impact Assessment (EIA) serves as a benchmark against which project progress in terms of environmental interventions may be quantified. Such periodic exercises provide insights on the part of primary as well as secondary stakeholders in the sense that strategies may be revised, goals redefined and praxes reformulated in the light of field-level lessons learnt during implementation of project objectives. Such a critical and heuristic manner of learning from past mistakes and making necessary amendments during the project cycle to aim at moving targets more often than not inform EIA studies that focus on the core issues of sustained environmentalism.

Environmental development is more often than not value-specific i.e. progress indicates changes at the level of value systems, broadbased worldviews, evolution of new managing agencies at the level of the mode of production and dynamic interfaces cohered among the stakeholders who happen to manage environmental affairs and the agenda of institutionalized joint participation mechanisms. Environmental management so emerges as a continuous exercise that should be able to reconcile contending role definitions between stakeholders and beneficiaries as well as resolve the tension of priorities between the notions of sustainability and environmental best practices.

Contributors to this Volume have tried to conceptualize the lessons learnt and identify the good practices of environmentalism so that future roadmaps for action research, advocacy and activism can be suitably cohered in the process. A range of experiences underpinned by scholarly analyses have been collected in this volume from different socio-economic as well as politico-cultural backgrounds that adequately reflect the state of environment and assess the quality of man-made and natural resources in the context of pluralist systems and multicultural societies. An understanding that repeatedly comes across during the perusal of the different academic papers published in this volume is the notion that social and physical environments impact upon one another and share an entire array of symbioses that have to be continuously investigated in order to explore indigenous solutions and establish convergence for the sake of ecological equilibrium.

Environment has both physical and human aspects, and the latter more often than not influence the former in terms of multiple strategies and image-building exercise. Environment - like any other limited resource in human society - has steadily emerged since the Industrial Revolution as an issue that assumes ownership of all the concerned stakeholders and - as an unfortunate result - is really nobody's ethical responsibility. So the North can afford to pollute mindlessly and then hold the South responsible for this sorry state of results while it is a fact that the patterns of consumption, industrialization and urbanization (development?) in the rich countries add more viciously to environmental degradation, global warming and climate change than elsewhere in the world. Not for nothing have we heard the doomsday conspiracy that the Third Word War would soon be fought on the issue of drinking water. Moral ecosophy has to be conscientiously practiced if we are to save Planet Earth from the apparently irreconcilable axes of power that have fractured the Global Economic Order and exposed the pretensions of the so-called process of Globalization that has only further deepened divides across our International Moral Order. But this is perhaps a noble aspiration without any conformation from our everyday politics of life and endangered realities.

Dr. Aminu Mamman
Dr. Chris Rees

# 2

# Understanding Development Policy Failures in Africa through the Lens of Management and Organizational Theories and Research

*Aminu Mamman*

*Christopher J. Rees*

**Keywords:** *Poverty Reduction Strategies; Implementation of Development Policies; Management and Organization theories.*

## Abstract

It is widely acknowledged by researchers and practitioners that development policies particularly in Africa have not delivered the desirable outcomes. This has been largely attributed to the inadequate attention paid to policy implementation. To address the issues of implementation, the article argues that organizational and management theories can advance our understanding of why good policies do not deliver desirable outcomes. Therefore, the paper outlines the current gap in the literature on development in general and poverty reduction strategy in particular. It highlighted that, because much of the research on Poverty Reduction Strategy (PRS) have been undertaken at institutional rather than at organizational level, the issue of implementation of poverty reduction interventions has been neglected. The article also demonstrates that researchers have not taken full advantage of the vast amount of literature in

**management and organization theory to unravel the complexity of implementing poverty reduction interventions. Therefore, to enable researchers to address the issue of implementation, a research proposal and an associated set of research questions are advanced in the paper to guide future research.**

## Introduction

Effective organisations are key to the eradication of poverty in Africa. Therefore, Poverty Reduction Strategies (PRS) should be associated closely and explicitly with organisational capacities to implement them (Blunt and Jones, 1992; Disch, 2004; Goldsworthy and Bennell, 2001; Jaeger and Kanungo, 1990; Kiggundu, 1991, 1993; World Bank 2005). In fact, investigations of Poverty Reduction Strategy Papers (PRSPs) in some African countries (Benin, Kenya, Malawi, Mali, Mozambique, Rwanda, Tanzania) reveal that organisational and institutional capacity issues are among the constraints impacting on the implementation of poverty reduction interventions (Bierschenk *et al.* 2003; Booth, 2003; Dante 2003; Evans and Ngalewa, 2003; Falck *et al.* 2003; Hanmer *et al.* 2003; Jenkins and Tsoka, 2003l Mutebi *et al.* 2003). Similarly, the Africa Commission Report (2005) recognised the need to build the capacity of African organisations. The conclusions of one recent study (Owusu, 2005: 157) specifically refer to organisational ineffectiveness in the public sector as one of the main reasons for the generally disappointing results of policy driven reform programmes in Africa:

> "The public sector in Africa is riddled with widespread ineffectiveness. Although some countries have implemented various reform programmes with the support of international development agencies, the results so far have been disappointing. One reason for the failure is that the policies have focused more on achieving macroeconomic stability than making the organisations effective."

Although more than three decades ago experts highlighted the existence of gap between policy formulation and policy implementation (see Grindle, 1980; Van Meter and Van Horn, 1975), researchers and policy makers are yet to address this anomaly adequately. UNCTAD (2002) pointed out that:

> "Progress in international efforts to alleviate poverty will depend as much on international development cooperation in resolving the problems associated with protectionism and debt as on the improvement of domestic policies institutions and governance in developing countries."

We will go as far as stating that progress in addressing poverty will depend as much on human, organisational and institutional capacity as on good governance. This is because successful implementation of good governance is dependent on the three dimensions of capacity (human, institutional, organisational).

The inadequate attention paid to systematic research on the capacity of poverty reduction organisations is a matter of concern. Intuitively, we know that effective organisations are critical to the success of poverty reduction programmes. However, there is little evidence on how this contribution is achieved or can be maximised. In contrast, in the private sector, such investigations have been conducted extensively. For example, despite methodological difficulties, extensive research has explored the impact of specific management practices on the achievement of organisational objectives (Connolly *et al.* 1980; Ostroff and Schmitt, 1993; Tusi, 1990). Similarly, extensive research has been conducted into how specific characteristics of organisations contribute to specific organisational outcomes (Sibbet, 1997). This has resulted in the development of body of knowledge on how to achieve organisational purpose and objectives (Daft, 2004; 2006; Huselid *et al.* 1997; Mamman, 2002a; Woodward, 1965). Beyond the issue of capacity, there is the issue of the willingness to use capacity and this is where organisational and human values play a significant role (Alas and Rees, 2006).

The objectives of this chapter are to:

- briefly review current research on PRS and outline issues pertaining to the failure of good policies on poverty eradication to produce desired outcomes;

- underscore the need for research on the capacity of poverty reduction organizations;
- discuss the potential contribution of management and organization theory can make in the study of the implementation of PRS;
- Provide direction for future research;

## Current Research on the Management of PRS

Much of the research on the management of PRS has been conducted at institutional level rather than at organizational level (Nayak *et al.* 2002). For example, attempts have been made to examine public expenditure management and poverty reduction interventions (Foster *et al.* 2002). In a nutshell, the bulk of literature on the management of PRS concentrate on the following: participation and ownership of PRS (e.g. UNCTAD, 2002; Booth *et al.*, 1998; ODI, 2002); public expenditure on PRS (e.g. Foster *et al.*, 2002; Foster and Zormelo, 2002; Foster and Mijumbi, 2002; Fozzard, 2001; 2002); monitoring and evaluation of PRS implementation (e.g. Hague 2000; McGee, 2000). Other investigations focus on the politics and government involvement in the preparation and implementation of PRS (e.g. Casson, 2001; Craig and Porter 2003; Gould and Ojanen, 2003; World Bank/IMF 2002; Piron and Norton 2004) and on the capacity to deliver PRS (Azeem and Ayamboo 2005). Another field of research on PRS focuses on the impact of PRS (e.g. Lentz, 2002; Booth, 2003; ODI, 2001). While these investigations have made a significant contribution to the implementation dimension of PRS, they have not directly investigated organizational capacities or their impact on the outcome of specific poverty reduction intervention. Therefore, there is the need for systematic independent academic investigation on the role and impact of the capacities of poverty reduction organisations on the achievement of specific poverty reduction objective.

So far, the body of knowledge on specific poverty reduction agencies and organizations is almost non existent. Yet in the last two decades or so, many African countries have established such organizations to help implement PRS. For example, Nigeria has established the National Directorate of Employment in order to address the problem of unemployment (Evawoma-Enuku and Mgbor, 2005). Under the poverty eradication strategy, the government has also established the Economic Empowerment Commission and the Poverty Eradication Commission. There are similar organizations across Africa. Understanding their capacity and internal processes will undoubtedly shed light on their effectiveness *vis-à-vis* PRS and poverty eradication effort in general.

## Issues Associated with Failure of Development Policies

It is now widely acknowledged that previous attempts to address poverty and underdevelopment in Africa through structural adjustment have not produced the outcomes expected (Aryeetey *et al.* 2000; Easterly, 2001). Similarly, the current use of aid to address development and poverty has been questioned and criticised (Azeem and Ayamboo; 2005; Booth, 2005; OECD DAC, 2003; Oxfam, 1999; 2000; Rogerson, 2005; Toye, 2003). Apart from the limitations of the policies themselves, the lack of effective organisations to implement the policies has contributed to their failure. At the best of times, the capacity of most African organisations is inadequate in terms of realising the objectives of current and previous policies on poverty. For example, market institutions are too under-developed to absorb the opportunities and slack produced by structural adjustment policies (UN 2002; Graham, 2002). The current PRS are not immune from the organisational problems that affected the implementation of previous development policies (Azeem and Ayamboo, 2005). Yet research effort to identify effective PRS and policies is not replicated in the management of poverty reduction organisations.

Evidence from Uganda indicates that the success of poverty reduction initiatives is much to do with commitment of officials located within the Ministry of Finance, Planning and Economic Development. Evidence also indicates that the staff of the ministries are well paid, and have developed a culture of high achievement openness and innovation (Foster *et al.*, 2002). Similarly Azeem and Ayamboo, (2005) reported that the successful implementation of poverty reduction programmes in Ghana was influenced by the effectiveness of the National Development Planning Commission. We believe that unless the *gap* between *macro* policy-oriented research and *micro* organisations-oriented research is filled, the issue of lack of impact of poverty reduction policies will persist. In other words, the problem of good policies without effective organisations to implement them will remain.

One of the disturbing revelations relevant to PRS has been the lack of adequate evidence between the increase in human capital and growth in per capita output (Easterly 2001; Roberts 2003). Synthesis of research by Easterly (2001) indicates that there is no apparent relationship between the growth in human capital and economic growth in Africa. These issues call for the need to look at other factors associated with the implementation of development policies. For example, given that development policies are implemented by organizations and people, it is logical to spend a significant amount of effort on investigating their role in the delivery of development interventions such as poverty reduction interventions.

## Need for Research on the Capacity of Poverty Reduction Organisations

There are a number of pressing reasons for studying poverty reduction organisations. First, as highlighted earlier, research on PRS in some African countries has revealed that organizational and institutional capacities are among the stumbling blocks in the implementation of PRS (Booth 2003). For example, in a study of the implementation of PRSP in Benin *et al.* (2003:69) observed that:

> "There were also concerns that capacity to implement the poverty strategy would be inadequate. The capacity problem was seen as the result of a number of factors, including unproductive use of staff time, poorly managed flows of funds and resources; and weak motivation linked to low staff salaries and high level of administrative corruption."

Similarly, Dante *et al.* (2003) points to the Malian Government's lack of capacity to adequately plan for the implementation of PRSP. Also Falck *et al.* (2003) underscored the lack of capacity to coordinate the implementation of PRSP in Mozambique. Organizational and institutional capacity to implement PRSP was also highlighted in Rwanda (Mutebi *et al.*, 2003) and Tanzania (Evans and Ngalewa, 2003). In spite of the evidence highlighted above, there is no systematic academic investigation of the capacities of poverty reduction organizations in Africa or anywhere else. Particularly, there is no systematic academic research on the impact of capacities on organizational ability to implement poverty reduction interventions and their outcomes. Therefore, research on these issues is long overdue.

Second, much of the research on the management of PRS has been conducted at institutional level (Nayak *et al.*, 2002). In a nutshell, the bulk of literature on the management of PRS concentrates on the following: public expenditure management and poverty reduction interventions (Foster *et al.*, 2002); participation and ownership of PRS (e.g. UNCTAD,2002; Booth *et al.* 1998; ODI, 2002); public expenditure on PRS (e.g. Foster *et al.* 2002; Foster and Zormelo, 2002; Foster and Mijumbi, 2002; Fozzard, 2001; 2002); monitoring and evaluation of PRS implementation (e.g. Hague 2000; McGee, 2000). Other investigations focus on the politics and government involvement in the preparation and implementation of PRS (e.g. Casson, 2001; Craig and Porter 2003; Gould and Ojanen, 2003; World Bank/IMF 2002; Piron and Norton 2004) and on the capacity to deliver PRS (Azeem and Ayamboo 2005). Another field of research on PRS focuses on the impact of PRS (e.g. Lentz, 2002; Booth, 2003; ODI, 2001). While these investigations have made a significant contribution to the implementation dimension of PRS, they have not directly investigated organisational capacities or their impact on the outcome of specific poverty reduction interventions. Therefore, there is an existent need for systematic independent academic investigation into the role and impact of the capacities of poverty reduction organisations on the achievement of specific poverty reduction objective.

Third, poverty reduction organisations have the mandate to implement, monitor, evaluate and report on the outcome of poverty reduction interventions. Therefore, their capacity to do so should be of acute interest to policy makers, donors, international development agencies and researchers.

Fourth, to appreciate the need for research on the capacity of poverty reduction organisations consider this: the production of PRSPs requires data to develop the strategy. However, many organisations dealing with the acquisition and management of data in African countries are under-resourced and poorly managed. Similarly, PRSP formulation requires the participation of NGOs and Civil Societies. But unless such organisations are properly managed and led, their contribution to the participatory process will be limited. In fact there is already criticism of the contribution and ownership of PRSPs (Azeem and Ayamdoo, 2005; Dante *et al.*, 2003).

Fifth, poverty reduction organisations are among the key building blocks of poverty reduction policy implementation. To ignore their internal management processes, capacity and interface within the wider policy framework is to negate the well established fact that the effectiveness of organisations is central to the social and economic development of any society (Bowie, 1991; Deyo, 1987; Gao, 1997). Thus, good policies without effective organisations to implement them represent little less than formalised dreams.

Sixth, the absence of effective organisations plays a role in limiting the impact of human capital development (Metcalfe and Rees, 2005). Unless more attention is paid to the effectiveness of organisations in all sectors of the economy, PRS are unlikely to succeed (Mamman and Sani, 2003; Mamman and Makonen 2003). Decades of research in management has revealed that specific organizational attributes such as strategy, structure and systems impact on employees' behaviour and organizational performance (Burns and Stalker, 1961; Chandler, 1962; Miles and Snow, 1978; Mamman, 1990; 1995; 1997; Mamman *et al.*, 1996; Mintzberg, 1983; Woodward, 1965). This body of knowledge has not been adequately utilised or indeed tested in development studies research (lewis, 2002).

Finally, there is no obvious body of knowledge on specific poverty reduction agencies and organisations. Yet in the last two decades or so, many African countries have established such organisations to help implement PRS. For example, Nigeria has established the National Directorate of Employment in order to address the problem of unemployment (Evawoma-Enuku and Mgbor, 2005). Under the poverty eradication strategy, the government has also established the Economic Empowerment Commission and the Poverty Eradication Commission. There are similar organisations across Africa. Understanding their capacity and internal processes will undoubtedly shed light on their effectiveness vis-à-vis PRS and poverty eradication efforts in general.

## Can Management and Organization Theory contribute?

While there are many factors that may account for the lack of direct relationship between human capital expenditure and economic development in Africa, we believe the absence of effective organisations to be one of the key causal factors. This is because empirical evidence has demonstrated that effective organisations are more able to create employment and also to pay relatively high wages (Abowd, 1990; Abrahamson, 1991; 1996; DiMaggio, and Powell, 1983). Also, evidence has shown that job creation and high wages are associated with the effective management of organisations (Daft, 2004; Huselid; 1995; Huselid *et al.*, 1997; Mamman, 1990; 1995; 1997; Mamman *et al.*, 1996). In fact more than three decades ago, management research has demonstrated that specific organizational attributes such as strategy, structure and systems impacts on employees' behaviour and organizational performance (Burns and Stalker, 1961; Chandler, 1962; Miles and Snow, 1978; Mintzberg, 1983; Woodward, 1965). Such body of knowledge is not adequately utilised in development studies research. The absence of effective organisations plays a role in limiting the impact of human capital development. Unless more attention is paid to the effectiveness of organisations in all sectors of the economy, PRS might not live up to their promise (Mamman and Sani, 2003; Mamman and Makonen 2003).

## Direction for Research

Based on the discussion so far, much of the problem associated with failure of poverty reduction initiatives is rooted in the absence of organizational capacity. Therefore, to address the existing gap in academic research on the capacity of poverty reduction organisations to implement specific poverty reduction interventions, researchers need to identify specific agency or agencies with a mandate to implement specific poverty reduction intervention. This is an approach widely used by management and organization theory researchers.

Researchers can contribute by utilising organisation and management theories and research to provide better understanding of the interface between PRS and the vehicles through which they are delivered. This can be achieved by undertaking an extensive review of *management* and *organisation* literature (in the private and public sectors) on the impact of capacities, structure, systems and processes on the achievement of organisational objectives. This will be followed by a comparison of findings from fieldwork with the literature generated.

## Research Questions

The main question that should guide researchers on the role of organizations in eradicating poverty is: What role do organisational and human capacities play in the implementation of PRS?

The specific questions that could guide the research could include:

— How does the capacity of a poverty reduction organisation influence its ability to implement specific poverty reduction intervention (e.g. to help the poor and the unemployed find a means of livelihood)?

— What are the specific levels of human and organisational capacities of the organization and its constituent's parts?

— Are there specific capacities unique to the implementation of poverty reduction? If so, what are they?
— How do the similarities and differences in capacities across organizations explain the nature and outcome of the implementation of PRS?
— What are the influences on the willingness of organisations and individuals to deploy their capacities to pursue poverty reduction objectives?
— What are the challenges and key organisational success factors for poverty reduction organisations in sub-Saharan Africa?
— What are the research and policy implications of the study?

## Methodology

In order to achieve the objectives outlined above, there is a need for researchers to define the key variables and how to measure them. This is necessary in order to provide the theoretical and practical comparative frameworks within which to study the role of organizations in the implementation of poverty reduction. While we recognize that the production of a universally acknowledged set of terms and terminology is likely to be a longitudinal and evolutionary process, we attempt to below to present definitions of key variables in this field.

## Organizations and Institutions

Richard Daft (2004) defined organization as "(1) social entities that (2) are goal-directed, (3) are designed as deliberately structured and coordinated activity systems, and (4) are linked to the external environment. In March 2003 DFID produced a guideline for "promoting institutional and organizational development. In the guideline, DFID describe organizations as groups of players who come together for a common purpose or to achieve specific objectives. They adapt their tactics and organization according to externally defined rules and regulations – the rules of the game. They play to, but are not the same as, those rules. For example, it is a rule of the game that football teams comprise a goal-keeper and ten outfield players; but the configuration, position and tactics of the outfield players are decided by the team as an organization. According to DFID are international, national or societal structures that guides and regulates individual and organizations' behaviour:

> "They may be formal – legal systems, property rights, enforcement mechanism; or informal – customs, traditions. They may operate at different levels – international (egg. WTO rules), national (e.g. Laws, constitutions), social (e.g. Norms of conduct, status of women), family (e.g. Inheritance rules). They may nest within larger institutions – e.g. Village-based collective institutions nested within the policy institutions of government. They often also referred to as "the rules of the game". The most widely used definition of Institutions is that used by North. They "consist of formal rules, informal constraints - - norms of behaviour, conventions, and self imposed codes of conduct – and their enforcement characteristics."

## Capacity

Using the World Bank's methodology, capacity can be viewed from three dimensions:

1. *Human capacity*. This refers to leadership and management skills and qualities possessed and used by people in poverty reduction organisations. It may also include skills and abilities of key staff to analyse poverty, design and implement strategies, policies, programs/projects, deliver services and monitor results pertaining to poverty.

2. *Organisational capacity*. This refers to groups of people with a clear purpose, objectives, and internal structures, processes, systems, staffing and resources to implement specific poverty reduction initiatives.

3. *Institutional capacity*: This refers to the formal rules and informal norms of using public resources and regulating public and private activities to provide the framework for implementing poverty reduction programmes and policies. Institutional capacity as defined above should not be within the scope of this research. However, the study should investigate the organization's/agency's understanding and capacity to apply norms and rules when using public resources.

## Effectiveness

The study should adopt the standard management and organisation theorists' definitions of effectiveness to explain the effectiveness of a specific poverty reduction organisation. Thus, effectiveness of the organisation should be viewed as the degree to which it is able to implement successfully economic empowerment interventions (Daft, 2004: 2006). It should also be viewed in terms of the organisation's ability to provide the services and support at the standard expected by its stakeholders (Daft, 2006). Stakeholders of poverty reduction organizations can include the following: Federal and State governments, job-seekers, current and passed beneficiaries of support, relevant NGOs, parents/relatives of the poor and job-seekers.

## Research Design

A good research design should have a structure that links the empirical data to be collected to the research questions (Blalock and Blalock, 1982; Robson, 1993). Studies may often adapt and use the well established and widely acknowledged *balance score-card* approach for evaluating organisational effectiveness and achievements (Kaplan and Norton, 1996). This methodology enables the evaluation of an organisation's key performance indicators and the factors (e.g. capacities) responsible for achieving them. The *balance community score-card* advocated by Holbeche (2006) can be adapted and used to assess the relationships between an organisation's specific stakeholders such as NGOs, federal and state governments, and beneficiaries of support. This approach can enable the collection of data that are directly relevant to answering research questions such as those outlined above. For example, how does an organization's capacity impact on its ability to successfully implement economic empowerment programme? What are the critical capacities necessary for the implementation of economic empowerment programme?.

## Data collection

Both quantitative and qualitative multiple case study approach as the primary data collection method can be used. The choice of case study methodology is derived from the need to understand complex social phenomenon such as organisational culture imbedded in poverty reduction organisations. This method will enable investigation of the phenomenon in context and provides the opportunity for deeper understanding of individuals, organisations and events (Krause, 1992). Also the use of multiple cases of implementation of poverty reduction initiatives offers the advantages of exploration, revelation of hidden features and, in-depth examination of multiple issues (Stoecker, 1991; Whipp, 1998). The use of multiple cases of implementation of poverty reduction will overcome the inherent risk of misjudgement because of the use of a single source of evidence (Hakim, 2000). The use of multi-site methodology also enables the establishment of trends which transcend the specifics of a single case (Yin, 1994).

The complexity of development environments, the contexts of policy implementation and the backgrounds of the informants are likely to demand the use of multiple tools and techniques to collect data (Jankowicz, 2000; Buchnan; Boddy and McCalman, 1988). For example, researchers may collect, examine and analyse policy statements, plans, and speeches to gain insights into implementation issues regarding poverty reduction intervention. They may collect documents (e.g. reports) relating to the organisation structure, systems achievement/ performance. Key staff and officers may be observed and interviewed in order to understand further the organisation's systems and processes. As part of the triangulation approach to the study, key stakeholders (e.g. beneficiaries of support, relevant NGOs, relevant federal and state departments) are among the likely sources of data and information.

## CONCLUSION

This article has outlined the current gap in the literature on development in general and poverty reduction strategy in particular. We have highlighted that much of the research on PRS have been undertaken at institutional rather than organizational level. We argue that the neglect of organizational issues is partly responsible for the failure of development policies to achieve desirable outcomes. The article also demonstrated that researchers have not taken full advantage of the vast amount of literature in management and organization theory to unravel the complexity of implementing poverty reduction interventions. It also noted the everlasting issue of failure of good development policies to deliver desirable outcomes due to the lack of effective mechanism and capacity to implement them.

We believe that studies which aid our understanding of the capacities of organisations in relation to the implementation of development interventions will make an important intellectual contribution to the field of development study in general and poverty reduction. Such studies will also contribute by bridging the current gap between *macro* policy-oriented research and *micro* organisation-oriented research, thus providing a more sophisticated understanding of why 'good' policy does not always produce the desired outcomes.

Finally, the field of management and organisational theory is well suited, as a discipline, to contribute to issues of organisational capacity and effectiveness (Lewis, 2002; Metcalfe and Rees, 2005). While certain section s of the sociologically driven international development academic community in the West may wish to eschew what they see to be capitalistic orientated business and management research and practices, we call for a more holistic approach to development management. Decades of research in the field of business and management have produced a significant body of knowledge on how to deal with issues pertaining to organisational capacity vis-à-vis implementation of policies and strategies. In fact, the current adoption of private-sector oriented management practices by public sector organisations and NGOs is a testament to the significant body of knowledge developed in the field of management. It is vital that this body of knowledge be explored and employed, particularly by international development policymakers and regulatory bodies, if development policy is to implemented effectively.

## REFERENCES

Alas, R. and Rees, C.J. (2006) Work-related Attitudes, Values and Radical Change in Post-Socialist Contexts: A Comparative Study. *Journal of Business Ethics*, 68(2): pp. 181-189.

Azeem, V.A. and Ayamboo, C. (2005). *Ghana: Experimenting with poverty and social impact assessments*. Paper commissioned by Eurodad: European Network on Debt and Development.

Bierschenk, T., Thioleron, E. and Bako-Arifari, N (2003). Benin. In David Booth, (ed.), *Fighting poverty in Africa: Are PRSPs making a difference?* London: Overseas Development Institute, 57-90.

Blunt, P. and Jones, M. (1992). *Managing organisations in Africa*. Berlin: Walter de Gruyter.

Booth, D (ed.), (2003). *Fighting poverty in Africa: Are PRSPs making a difference?* London: Overseas Development Institute.

Booth, D., Holland, J., Hentschel, J., Lanjouw, P. and Herbet, A. (1998) *Participation and Combined Measures in African Poverty Assessment: Renewing the Agenda* DFID Social Development Division. Africa Division: February 1998.

Bowie, A (1991). *Crossing the industrial divide: State, Society and the politics of economic transformation in Malaysia*; New York: Columbia University Press.

Brown, S.L and Eisenhardt, K.M (1997). The art of continuous change: Linking complexity theory and timed-paced evolution in relentlessly shifting organisations. *Administrative Science Quarterly*, 42(1), pp. 1-34

Buchanan, D., Boddy, D. and McCalman, J. (1988). 'Getting in, Getting on, Getting out, and Getting Back'. In: Bryman, A (ed.) *Doing research in organisations*. London: Routledge; pp. 53-67.

Burns, T. and Stalker, G.M. (1961). *The Management of Innovation*. London: Tavistock.

Casson, K. (2001) *Governance and the PRSP process: A Review of 23 IPRSPs / PRSPs*, Governance Department, DFID.

Chandler, A.D. (1962). *Strategy and Structure*. Cambridge; Mass: M.I.T Press.

Christiansen, K. and Hovland, L. (2003). *The PRSP initiative: Multi-lateral policy change and the role of research*. ODI working paper No. 216.

Connolly, T. Conlon, E.J. and Deutsch, S. J. (1980). Organisational effectiveness: A multiple-constituency Approach. *Academy of Management Review*, 5, pp. 211-217.

Crompton, R. and Jones, G. (1988). 'Researching white-collar organisations: Why sociologists should not stop doing case studies'.

In: A. Bryman (ed.) *Doing Research in Organisations*. London: Routledge; pp. 68-81.

Craig, D. and Porter, D. (2003) Poverty Reduction Strategy Papers, a new convergence, *World Development Report* Vol.31, No. 1: pp. 53-69.

Daft, R. (2004) *Organisation theory and design*. London: Thomson.

Daft, R. (2006) *The new era of management*. London: Thomson.

Dante, I., Gautier, J., Marouani, M. and Raffinot, M. (2003). Mali. In: David Booth, (ed.), *Fighting poverty in Africa: Are PRSPs making a difference?* London: Overseas Development Institute, 153-186.

Deyo, F. (1987). *The political economy of the New Asian industrialism*. Itacha: Cornell University Press.

DFID (2003). *Promoting Institutional and Organizational Development*. Department for International Development, London: United Kingdom.

Disch, A. (2004) *An Independent Review of World Bank Support to Capacity Building in Africa: The Case of Malawi*. Background Paper, OED evaluation of World Bank support for capacity building in Africa. Washington, DC: World Bank.

Evans, A and Ngalewa, E (2003). Tanzania. In: David Booth (ed.), *Fighting poverty in Africa: Are PRSPs making a difference?* London: Overseas Development Institute, 247-273.

Evawoma-Enuku, U. and Mgbor, M. (2005). The national directorate of employment's open apprenticeship scheme in Nigeria: New wine in old winskin? *Education and Training, 47(4), pp. 325-336.*

Falck, H., Landfald, K. and Rebelo, P. (2003). Mozambique. In: David Booth, (ed.), *Fighting poverty in Africa: Are PRSPs making a difference?* London: Overseas Development Institute, 187-216.

Foster, M., Fozzard, A., Naschold, F. and Conway, Y. (2002). How, when and why does poverty get budget priority. ODI working paper #/68.

Foster, M. and Zormelo,D (2002) *How, When and Why does Poverty Reduction get Budget Priority? Ghana Case Study*. ODI Working Paper, No. 164: London, April.

Fozzard, A. (2001) *The Basic Budgeting Problem: Approaches to Resource Allocation in the Public Sector and their Implications for Pro-poor Budgeting*. ODI Working Paper, No 147: London: July 2001.

Fozzard, A. (2002) *How, When and Why does Poverty Reduction get Budget Priority? Mozambique Case Study*. ODI Working Paper, No. 167: London May.

Gao, B. (1997). *Economic Ideology and Japanese Industrial Policy: Developmentalism from 1931 to 1965*. Cambridge: Cambridge University Press.

Goldsworthy, P. and Bennell. P. (2001). *"African Capacity Building Foundation: Performance Evaluation Mission."* Report to Department for International Development (DFID). London.

Gould, J. and Ojanen, J. (2003). *Merging in the circle: the politics of Tanzania's Poverty Reduction Strategy*. Policy Papers 2/2003; University of Helsinki, Institute of Development Studies

Grindle, M.S. (1980). Policy Content and Context in implementation. In Grindle, M. (ed.) *Politics, and policy implementation in the third world*. Princeton: Prince University Press, pp. 3-34.

Gummesson, E. (2000). *Qualitative methods in Management research*. London: Sage Publications.

Hakim, C. (2002). *Research design*. London: Routledge.

Holbeche, L. (2006). *Understanding change: Theory, implementation and success*. London: Butterworth.

Hanmer, L., Ikiara, G., Eberlei, W. and Abong, C. (2003). Kenya. In: David Booth, (ed.), *Fighting poverty in Africa: Are PRSPs making a difference*? London: Overseas Development Institute, 91-118.

Hauge, A. (2000) 'Strengthening Capacity for Monitoring and Evaluation in Uganda, a Results Based Management Perspective'.

Huselid, M.A., Jackson, S.E. and Schuler, R.S. (1997). Technical and strategic Human Resource Management effectiveness as determinants of firm performance. *Academy of Management Journal*, 40: pp. 171-188.

Jaeger, A.M. and Kanungo, R.N. (1990). *Management in developing countries*. London: Routledge.

Jankowicz, A.D. (2000). *Business research projects*. London: Thomson Learning.

Jenkins, R and Tsoka, M (2003). Malawi. In: David Booth, (ed.), *Fighting poverty in Africa: Are PRSPs making a difference*? London: Overseas Development Institute, 119-152.

Kaplan, R. and Norton, D. (1996). *The balance scorecard: Translating strategy into action*. Boston: Harvard University press.

Kinggundu, M. (1991) The challenges of management development in sub-saharan Africa. *Journal of Management Development,* 10(6), pp. 32-47.

Kinggundu, M. (1993) *Managing organisations in developing countries.* Kumari Press, Connecticut

Krause, P. (1992). *The Battle for Homesteads, 1880-1892*. USA: University of Pittsburg Press .

Lentz, A. (2002) *Assessing the impact of Uganda's Poverty Action Fund: a participatory rural appraisal in Kamuli district,* New York: World Learning for International Development.

Lewis, D. (2002) Organisation and management in the third sector: towards a cross-cultural agenda. *Non-Profit Leadership and Management,* 13(1): pp. 67-83.

Mamman, A. (1990). Employee preferences for pay systems: Theoretical approaches and an empirical test. *International Journal of Human Resource Management*, vol. 1 (3), pp. 329-340.

Mamman, A. (1995). Distinguishing the purpose and objectives of new working arrangements. *Australia and New Zealand Academy of Management Journal,* 1(1), pp. 27-43.

Mamman, A., Sulaiman, M. and Fadel, F. A. (1996). Exploring attitudes to pay systems within and across cultures. *International Journal of Human Resource Management,* 7(1) pp. 101-121.

Mamman, A. (1997). Employees' attitudes to pay systems. *Journal of Social Psychology* 137(1), pp. 33-42.

Mamman, A. and Saffu, K. (1998). Shotermism, Control, Quick-fix and Bottom line: Towards Explaining the Anglo-saxon Approach to Management. *Journal of Managerial Psychology,* 13 (5/6), pp. 291-308.

Mamman, A. (2002a). Adoption and Modification of Management Ideas in Organisations: Towards an analytical framework. *Strategic Change*; 11; pp. 379-2002.

Mamman, A. (2002b). Managerial Attitudes to Government intervention in Malaysia. *Asia Pacific Business Review,* 9(1), pp. 1-20.

Mamman, A. and Sani, B. (2003). *Skill Needs of Small and Medium-Sized Enterprises: An investigation of the interface between labour market and vocational educational training in Nigeria*. Report to the Department for International Development, UK.

Mamman, A. and Makonen, S. (2003). Human Resource Management in Ethiopia. In: K. Kamoche, *et al.* (eds.) *Human Resource Management In Africa*. London: Routledge, pp. 103-115.

Metcalfe, D. and Rees, C.J. (2005) Theorising Advances in International HRD. *Human Resource Development International*, 8(4): 449-465.

McGee, R. (2000) 'Analysis of Participatory Poverty Assessment and Household Survey Findings on Poverty Trends in Uganda, Mission Report', 10-18th February, 2000.

Miles, R. and Snow, C (1978) *Organisational Strategy, Structure and Process*. New York: McGraw-Hill.

Mintzberg, H. (1983). *Structures in Fives: Designing effective organisations*. London: Prentice-Hall.

Mutebi, F.G., Stone, S and Thin, N (2003). Rwanda. In: David Booth, (ed.), *Fighting poverty in Africa: Are PRSPs Making a Difference?* London: Overseas Development Institute, 217-246.

Nayak, R., Saxena, N.C. and Farrington, J. (2002). *Reaching the poor: influence of policy and admin processes on the implementation of Government poverty schemes in India*. ODI working paper No. 175.

ODI (2001) *Key Findings on PRSPs to Date*. DFID PRSP Monitoring and Synthesis Project Synthesis Note No. 1 : September.

ODI (2002) *Assessing Participation in PRSPs in sub-Saharan Africa* (report on IDS study). DFID PRSP Monitoring and Synthesis Project Synthesis Note No.3: February.

Ostroff, C. and Schmitt, N. (1993). Configuration of organisational effectiveness and efficiency. *Academy of Management Journal*, 36; pp. 1345-1361.

Owusu, F. (2005) Livelihood strategies and performance of Ghana's health and education sectors: exploring the connections. *Public Administration and Development*, 25(2): pp. 157-174.

Piron, L. and Norton, A. (2004). *Politics and the PRSP approach: Uganda Case Study*. ODI Working Paper no. 240.

Sibbet, D. (1997). *Harvard Business Review, 75 years of Management Ideas and Practice 1922-1997*. Boston: *Harvard Business Review*.

Stoecker, R. (1991). Evaluating and rethinking the case study. *Sociological Review*, 39(1), pp. 88-112.

Tusi, A. (1990). A multiple-constituency model of effectiveness: An empirical examination at the human resource subunit level. *Administrative Science Quarterly*, 35, pp. 458-483.

UNCTAD (2002). *The least developed countries report: Escaping the Poverty Trap*. New York: United Nations.

Van Meter, D. and Van Horn, C. (1975). The policy implementation process: A conceptual framework. *Administration and Society,* 6(4), pp. 446.

Whipp, R (1998). 'Qualitative methods: Techniques or size?' In: K. Whitfield and G. Strauss (eds.), *Researching the world of work: Strategies and methods in studying Industrial Relations.* ILR Press, New York, pp. 51-63.

Woodward, J (1965). *Industrial organisation: Theory and practice.* London: Oxford University Press.

World Bank/IMF (2002) *PRSP Comprehensive Review*. Washington, DC: World Bank and IMF.

World Bank (2005). *Capacity Building in Africa: An OED Evaluation of World Bank Support*. Washington D.C.: World Bank.

# 3

# Flood and Riverbank Erosion: An Estimation of Economic Losses in Malda District, West Bengal

*Arunayan Sharma*

## ABSTRACT

The Malda district in West Bengal is a highly flood and riverbank erosion affected area of India has been witnessing devastating flood and riverbank erosion since last three decades. The flood in the district mainly caused by the high water level rise in the river Ganga, Mahananda, Fulhar, and Kalindri and Tangon, which are flowing through the district. Out of fifteen blocks of Malda district Kaliachak-1, Kaliachak-2, Kaliachak-3, Manikchak, Ratua-1, and some portion of English Bazar are the most vulnerable zone in respect of flood and riverbank erosion. The nature and factors leading to riverbank erosion problem are much more complicated to allow instant fixes. To refuel the riverbank erosion protective measures are continuing in the form of construction of earthen embankments and spurs along the Ganga riverside in the name of river training and flood control measures. Since last three decades the devastating effect of flood and riverbank erosion went a space washing out of spurs, refined embankments in tandem and the fragile measures taken so far failed to secure impact on the rate of riverbank erosion and in effect led to loosing more geographical and political areas of the district.

Periodicity of flood and riverbank erosion continuing unabated has thrown the existence of many habitat into an unworthy situations. A provisional estimate of losses due to flood and riverbank erosion in terms of Economic Loss has been ascertained on the basis of previous works and reports received so far from different Government departments, offices of the district administration and that of from the extensive field survey done by the Centre for Ecological Engineering in the various flood and river bank erosion affected areas of Malda.

**Keywords:** *Flood, Riverbank Erosion, Ecological Loss, Geographical and Political Area Loss, Political Boundary Dispute, Economic Loss, Property Loss, and Livelihood Loss.*

## Introduction

The recurring flood in the Ganga and its bank erosion in Malda in West Bengal are of very serious concern since last three decades. Since last three decades around 269 sq. km of Malda's land eroded owing to flood and riverbank erosion which includes around 288 villages of Kaliachak – 1, Kaliachak – 2, Kaliachak – 3, Ratua – 1, Manikchak and English Bazar block. Around 25 lakh people were directly or indirectly affected by recurring flood and riverbank erosion, turning around 2 lakh Landless Flood Affected Refugee and add to more around 1.2 lakh people lost their livelihood, settlements and prime agricultural lands (Anon., 2003, Sharma, 2000, 2006).

The main flood and riverbank erosion threat to this region is from the river Ganga. The mighty river Ganga in its meandering belt is slowly shifting its course of flow direction tangentially inside Malda district. By this meandering phenomenon, the river erodes the lands of Malda district, that of West Bengal and deposits soil on it's opposite riverbank side adjacent to the State of Bihar and Jharkhand. During the past three decades few new islands amidst the river Ganga have been formed. Most of the Landless Flood Affected Refugees settled in those islands to sustain their lives and their families.

They are now living their as isolated islanders. Due to some political reason these islands are disputed lands. Neither the West Bengal government nor Bihar and Jharkhand governments recognised them as their citizens. No voter cards, no ration cards were issued to them. No birth certificates for their children are provided to them by any government (Sharma, 2001, 2002, 2004).

The flood and river bank erosion problem in Malda district is so severe that, if in future the Ganga changes her course of flow along the either end of the district, the total people of Malda would also literally become Landless Flood Affected Refugees. The background of this citizenship problem is that, it would inevitably follow the decision implied in the footnote of the map published by the Survey of India (Map No. 72 P/ 13-1946) in 1946, where it is stated that the line demarcation between West Bengal and Bihar is ascertained on the basis of the deep channel course of the Ganga. It was also mentioned that in future the changes in the course of direction of the Ganga would affect this line of demarcation. Consequently due to this dispute and grave situation every year the inhabitants of Malda district are losing their valuable lands (Sharma, 2001,2004, 2005a, 2006b, Rudra –2002, 2004).

To cope up with or to come out of this recurring flood and bank erosion threat, many earthen embankments built so far by the State Irrigation Department and Farakka Barrage Authority for river training and flood control measures proved to be useless and futile following wrongly drawn ineffective plans. A large portion of Malda's land is under continuous flood and riverbank erosion. The main threat is from the river Ganga which doing havoc both in its natural environment and its socio-economic structure (Sharma, 2001, 2006b, Rudra, 2002).

## Location and Study Area

Among the different districts of West Bengal, Malda is the most important district, which is, situated almost middle part of the state. Malda district connects North Bengal With South Bengal by road and railway The Malda district in

West Bengal as a whole has been chosen as the study area. The Malda district is located under the Jalpaiguri Division of West Bengal covers an area of 3713 km, is situated between the latitudes 24°40'20" N and 25°33'08" N and longitude 87°45'50" E and 88°28'10" E in the Northern Hemisphere of the Globe. The district stands between Darjeeling in the North and the city Kolkata (=Calcutta) in the South. The district is demarcated by the international boundary of Bangladesh in the East. The main river of the district, the Ganga forms the Southwestern boundary and Dinajpurs in the North and Murshidabad in the South (Anon., 1998, Ghosh, 2004, Lambourn, 1918, Sengupta, 1969).

## General Configuration of Malda District

On the basis of topography and drainage pattern, Malda district divides into three parts. These are *Tal, Diara* and *Barind.* Malda is a low-lying plain slopping gently towards southern directions in which river flows. Diara region is more fertile than the other.

### Barind

The barind tract of Malda district is of comparatively big lands which lies on the east of Mahananda River. The characteristic physical feature of this block is wide undulation with successive ridges and depressions with small watercourses in the valleys and with small patches of sal forests. Drinking water becomes scarce during hot weather. Old Malda, Gazol, Bamongola and Habibpur Police stations are present within this tract.

### Diara

The Diara consists of a strip along the west and south sides of the district. Its formation is the result of fluvial action by the Ganga mainly depositional action in the latter Pleistocene age. Silts of varying sizes and sands are dominant composition of soil. English Bazar, Kaliachak, Manikchak are the main police station found under this tract.

## Tal

The Tal region is situated to the west of Mahananda River. It is a low-lying part and is subjected to inundation with the actions of the rivers, mainly of Ganga, Mahananda and Kalindri. Construction of protection embankments has reduced the chances of flood in this physical division. The Tal area gradually slopes down towards the south and it merges with the Diara region towards west (Anon., 1998, Ghosh – 2004, Lambourn – 1918, Sengupta – 1969).

## Geology of Malda district

Malda district is covered by alluvium. Most the area to the east of Mahananda river is occupied by an old alluvium formation known as Barind the old alluvium formation is considered to have formed in the Pleistocene period. West of Mahananda is occupied by recent alluviums consisting of sandy clay and sand along the course of river and fine silt consolidating into clay in flatten parts of the plain (Anon., 1998, Sengupta, 1969).

## Soil of Malda district

The district having different physical and physiographical characteristics is covered by alluvium of two different ages. Older alluvium dominates the baring region while newer alluvium dominates both Tal and Diara region. The southern portion of the district receives Ganga silt is the most fertile and formation of Diara is the result of fluvial action by the Ganga. Poor soil found in Barind region (Anon., 1998, Ghosh, 2004).

## Drainage Pattern of Malda District

Ganga and Mahananda, two important river flow through the district. Manananda is the main river originated from sub - Himalayan region and flow southerly direction and has divided the district into two parts. Kalindri is an important tributary of Mahananda. Kalindri divided the western part of Malda district into two parts. Northern half is low-lying area, which slopes towards the central portion of the district known as Tal regions. The southern half is most fertile and known as Diara the rivers are perennial almost throughout the year but in summer season Mahananda shrink to rivulets. River Punarbhaba is flowing along the extreme eastern most part of Habibpur and Bamongala and enters Bangladesh. River Gomani has been poured into Ganga near Farakka Barrage. River Fulhar is the important tributary of Ganga and has joined Ganga near Bhutni char. The confluence of Mahananda and Kalindri has been observed near Nimasari of English Bazar. River Pagla is the most important river of the district (Anon., 1998,Ghosh, 2004, Lambourn, 1918, Sengupta, 1969).

## Causes of Flood and River Bank Erosion in Malda

Ganga River in Malda always tries to maintain a dynamic equilibrium following the mass, momentum and energy to a fluid flow system accomplished by its self-controlled volume. But after the completion of the Farakka Barrage between Malda and Murshidabad district above the river Ganga in 1971, these dynamic equilibrium has been interrupted in the natural flow of river Ganga increasing the tendency of erosive power which has in turn aggravated the rate of bank erosion, siltation and flood in the Up Stream of Malda district. It is also found that the Malda district was subject to severe flood and riverbank erosion before the construction of Farakka Barrage (Chakrbarti – 1998, Sengupta – 1969). The river Ganga was already started shifting its course of flow in the Up Stream of Malda district from Farakka Barrage (Sharma, 2003, Rudra, 2000). It has been found by the critical observation that in near future the river Ganga will divide its course keeping the Farakka Barrage aside and possibly this path will be either through Kalindri or Choto Bhagirathi or Pagla- the almost moribund rivers in Malda district (Banerjee and Chakraborty – 1983, Sharma – 2003, Rudra – 2002).

The Ganga on the down stream of Maharajpur in Bihar flows on the right side of Bhutnidiara Island and then enters this district of Malda. The deep channel more or less hugging the bank line between Manikchak ghat to Panchanandapur mainly .The problem of erosion along the Left Bank of river Ganga at Up Stream of Farakka Barrage in the Malda has been of serious concern since early sixties, but the problem manifested it self to forcible magnitude during post Farakka condition. Since 1937 when the reach between Rajmahal and Farakka was straight the river started meandering towards the left bank and in 1963 the village Panchanandapur about 20km upstream of Farakka Barrage was under the severe threat of river bank erosion (Anon. – 2004, Chakrabarti – 1998, Ghosh and Bandhopadhay – 1986, Sharma – 2002, 2004, Rudra – 2000, 2004).

The area of Ganga-Bramhaputra basin is 1,50,000 sq km. In a depressed Synclinal shield region the accumulation of silt and clay during a prolonged period of 70 lakhs years such a basin plain was formed. Geologists named this basin - 'Bengal Basin'. Ganga and Brahmaputra River bring an amount of 166.70 crore-ton silt in a year and deposit them in a fashion of layer over layers and ultimate configuration was attained. In this big basin region each sq. km. area receives 10.30 lakh cubic mt water through rainfall in a year. Taking the bed of Ganga and Bramhaputra such a huge amount of water passes and reach to the Bay of Bengal. Keeping Meghalaya plateau at the eastern and Chottanagpur shield at the western part the narrow passage of this Basin consists a length of 200 km (approx) and allow the ice molten water of the rivers of Uttar Pradesh, Nepal and Tibet to pass away. It is known as Rajmahal-Maghalaya gap. Through this gap the enormous quantity of water (85,000 crore cubic mt) of the Brarmhaputra, Tista, Mahananda and other rivers of North Bengal takes their way. But as the rivers can't contain such a huge quantity of water in their shallow beds. Occurrences of flood and that of the riverbank erosion are very common in Malda district. Parua – 2002, CEE – 2004, Sharma – 2005a, Rudra – 2000b).

Stream does more than shift sediment by repeated scour and fill along the bed. The actively erode by Channel deepening or by down cutting of the streambed. Channel widening through bank curving or undercutting, and Channel extension, that is headward of regressive erosion by streams and gullies. This erosion is of two basic types in the case of river Ganga. Firstly, the valley floor over plain of Bengal consists of river laid sediments of alluvium where the river shift her channel and readjusts her bed by a little erosion and deposition. As a new channel is formed, an older channel is abandoned and filled, or as one bank of the meandering mighty Ganga collapses, the river shifts in that direction and accretion follows on the opposite bank. Such erosion in alluvium is fundamental in the development of alluvial plains, being a mechanism whereby the Ganga makes short or long-term adjustments. Secondly, streams cut their way into upland where they usually erode into compact rocks (Anon.- 2004, Chakrabarti – 1998, Parua – 1999, Pawar – 2006, Sen – 2003,Singh – 2003, Subramanya – 1998, Sharma – 2005b).

Cutting in the established channel of Ganga involves bed erosion, bank undermining, and bank collapse. Erosion of the alluvium on the riverbed involves lifting and pushing of loose particles, particularly by turbulent waters. Bank collapse also as the river banks of the mighty Ganga are under cut. At some critical angle the bank slope becomes unstable, and the upper part of the bank falls in or slides down, producing a gentler slope until undermining begins a new. In the case of river Ganga bank collapse is also aided by groundwater seepage or effluent seepage towards the river in the dry months of the year i.e. from March to August. Pipelines of natural origin, which bring the water flow of groundwater by seepage, make the bank sidewalls weak and weak and there by create big fissures and faults and bigger knoches, which in future will aid bank slumping by striking of the river current (Parua – 2002, 2003, Pawar – 2006, Sen – 2003, Singh - 2003).

Where riverbanks consist of cohesive silt and clay and where a high proportion of the load is carried in suspension, channels maintain steep banks and deep channels. However, silty bank lose their cohesiveness as they become saturated at times of high water; as long as the water level remains high, hydrostatic pressure may support the bank; however, when the water drops the bank may cave in. On the other hand loose sands or gravels tend to fall into a stream, maintaining steep banks that are quite unstable. Erosion is rapid and great quantities of bed lead are immediately available. This favour channel widening instead of channel deepening. Excessive bank slumping with an aid of water saturated block collapsation along bank lines is the main reason of bank erosion in Malda consequent upon river course changes (Anon. – 2004, Chakrabarti – 1998, CEE – 2004, Sing et.al – 1980, Sharma – 2005a, 2006).

By the action of River Ganga and flood the environmental changes and economic losses are taking place in Malda district during every flooding season. Deposition of slits increasing the river Ganga bed height at this place. Little increase in the water level at this critical point by the glacier fed river Ganga overflows and creates flood. It was also observed that being low altitude area the lower portion of the district known as *Diara* is situated in the Indo-Gangetic flood plains along the course of the river Ganga is most floods prone. Where as the upper part of the district situated in the Sub - Himalayan ranges known as *Tal* and its western part known, as Barind is less flood affected. Careful observation of the behaviour of the river Ganga at this critical point reveals that the river Ganga has been oscillating about north-south axis between Rajmahal and Lalgola with lateral bank erosion and complementary bar building. The river Ganga which was more or less straight in this sector during 1922-1936, has developed convexity eastward, upstream of Farakka Barrage and westward downstream of it. The eastward convexity upstream of Farakka Barrage is related to the development of a point bar between Rajmahal and Begamganj on the right bank and retreat of the left bank in the Panchanandapur-Taipy-Khejuria sector. Similarly the westward convexity downstream of Farakka Barrage is related to a bar development in the left bank and retreat of the right bank by lateral bank erosion in the Farakka-Dhulian and Nurpur-Lalgola sectors (Anon. – 2004,

Chakrabarti – 1998, Ghosh and Bandhopadhyay – 1986, Parua – 1999, Sharma – 2003, 2005b, Vendre - 2002).

## Methodology

Due to the recurring flood and riverbank erosion the landscape of Malda is changing every year, so the ecological and economic balances of the district. Every year the geographical area of Malda that of West Bengal is reducing and its political boundary under dispute. Still no such measures were taken to tackle such problems. Apart from its land losses a large amount of money also wastage in the name of construction of earthen embankments and spurs. After and during flood a huge amount also being spent on rescue and relief operations and to restore flood damages. Thousands of people loose their lands both in respect of settlements and agricultural lands and livelihoods. From the point of view studying such Ecological and Economic losses is very difficult. The Centre for Ecological Engineering to estimate such Economic Losses has taken an attempt. A wide range of data's was used in this paper collected from the administration. But this study is mainly based on the field survey done by the Centre for Ecological Engineering in different flood and riverbank erosion affected areas on Malda district.

## Estimation of Economic Losses

New islands at the middle of the river Ganga also emerged but this economic and fertile land losses are not easily compensated by growth of many low laying areas which are also susceptible to regular annual flooding. Annual flood not only responsible for havoc loss of human lives, livestock's, properties and prime agricultural lands but affects lakhs of people, making them landless refugees. Once the land eroded, engulfed by the river Ganga it never comes back. The river Ganga continuous cutting its bank one side and deposits silts on its opposite bank. The political boundary between Bihar, Jharkhand and West Bengal indicated by the river

Ganga. While the river Ganga cutting its embankments in Malda District, West Bengal, in the same time creating new islands and sandbanks adjacent in Bihar and Jharkhand States. People those who are loosing their settlements and prime agricultural lands can not allow to be resettle on these newly formed islands or on sandbanks due to political reason. Slowly during every recurring flooding season West Bengal loosing its geographical area. It is one of the major social problems in Malda District, which is also a great political debate (Anon.- 2004, Sharma – 2005a, 2005b, Mitra - 2004). It has been estimated that during past three decades as many as 288 villages in this district have totally gone out of existence due to river bank erosion and the people have to shift in other places from their original homestead. Again total number of families affected is about 62,230, and population affected 25,51,00. The total loss in the district area is about 269 sq. km which cost around 7062 crores.

**Table 3.1. Economic Losses in different sector due to Flood and Riverbank Erosion in Malda district, West Bengal**

| *Sectors* | *Unit area / No. Affected* | *Cost of Damage (in Rs. Crores) (Damage in Crores)* |
|---|---|---|
| Roads and Bridges | 570 km | 140 |
| Agricultural land | 110 sq. km | 760 |
| Homes/Buildings, | 67,000 Nos | 1230 |
| Crop damage | 2200 sq. km | 265 |
| Livestock | 70,000 Nos | 105 |
| Relief and Rescue measures | 935 batch | 225 |
| Handloom Sector | 65 Nos | 165 |
| Agricultural Sector | 650 sq. km | 120 |
| Embankment and Spur (Protecting Measures) | Every year up to 10th retired embankment and 20 spurs | 460 |
| Fish Farming | 1100 sq. km | 220 |
| P.H.E. Restoration work | Up to the year 2006 | 125 |

| Sectors | Unit area / No. Affected | Cost of Damage (in Rs. Crores) (Damage in Crores) |
|---|---|---|
| Electrical Board | Up to the year 2006 | 27 |
| Sericulture | Up to the year 2006 | 38 |
| School | 65 Nos | 31 |
| Small Scale Industries | 330 Nos | 23 |
| Food and Supply | Up to the year 2006 | 68 |
| Horticulture | Up to the year 2006 | 45 |
| Panchayat property | Up to the year 2006 | 80 |
| Health Sector | Up to the year 2006 | 290 |
| Government property | Up to the year 2006 | 415 |
| Private property | Up to the year 2006 | 2290 |
| Total Economic Loss | Up to the Year 2006 | Rs.7062 |

**Table 3.2. Showing Settlements and Population affected due to Flood and Riverbank Erosion in Malda district, West Bengal**

| Name of Block | No. of G.P. | No. of Village | No. of Family | No. of Population |
|---|---|---|---|---|
| Kaliachak -1 | 32 | 54 | 14,500 | 63,500 |
| Kaliachak- 2 | 47 | 67 | 16,750 | 75,600 |
| Kaliachak- 3 | 51 | 61 | 10,500 | 54,500 |
| Ratua- 1 | 26 | 42 | 6,250 | 25,000 |
| Manikchak | 30 | 45 | 9,010 | 14,000 |
| English Bazar | 15 | 19 | 5,220 | 22,500 |
| Total | 201 | 288 | 62,230 | 2,55.100 |

**Table 3.3. Various categories of land eroded in Malda district, West Bengal.**

| Category of lands | Area Eroded in s q. km. |
|---|---|
| Agricultural land | 92.00 |
| Orchards | 37.00 |
| Settlements | 105.00 |
| Wetlands | 16.00 |
| Others | 19.00 |
| Total | 269.00 |

**Table 3.4. Showing crops under thereat due to flood and riverbank erosion in Malda district, West Bengal.**

| *Blocks* | *Crops under threat* |
|---|---|
| Ratua -1 | Aman, Aus, Boro, Wheat, Vegetables, Oil seed |
| English Bazar | Aus, Pulses, Aman, Wheat, Mustard, Boro, Mango, Sugarcane |
| Manikchak | Pluses, Maize, Wheat, Sugarcane, Oilseed, Mango |
| Kaliachak-1 | Sugarcane, Rice, Maize, Jute, Wheat, Pulses |
| Kaliachak-2 | Maize, Jute, Mango, Sugarcane, Wheat |
| Kaliachak-3 | Oil seeds, Vegetables, Mango, Pulses |

**Table 3.5. Showing number of family and people turned Landless Flood Affected Refugees in Malda district, West Bengal**

| *Blocks* | *Family* | *Population in thousands* |
|---|---|---|
| Kaliachak – 1 | 6,250 | 36,000 |
| Kaliachak-2 | 7,500 | 35,500 |
| Kaliachak-3 | 6.100 | 28,750 |
| Ratua-1 | 1,200 | 52,500 |
| Manikchak | 6,000 | 29.000 |
| EnglishBazar | 1,800 | 11.000 |
| Total | 28,850 | 1,92,750 |

## Discussion

Flood in district Malda, West Bengal being a disaster has given ruinous results by its occurrence in every year inundating large of small areas, damaging crops and other constructional items, private or public. Due to the shifting nature of river Ganga and that of flood and river bank erosion creating enormous mental and economic pressure on the local people and problems in every sphere primarily from agricultural sector to settlements and others. Mango orchards play a vital role in the economy of this district. Due to heavy flood in the rural areas, these mango orchards are in spate.

Sericulture, which constitutes a considerable part of the district's economy, has also been partly ruined by bank erosion and flood. The populations, who are mainly dependent on agricultural activities, are trying to change their cropping pattern and crop combination method. The studied problems are variegated, one is linked with the other and one creates the other problem. Every year crores of rupees are being spent to prevent flood and post flood situation. The total economic cost caused by flood and bank erosion and allied activities is increasing year after year. The quantitative and qualitative analysis of flood damage and flood loss will prove that at present it is not possible to stop the occurrence of flood, but it can somewhat be checked by taking some measures (Sharma, 2001, 2003a, 2004).

According to its wetlands wealth, the district Malda in West Bengal an exceptional one. As per the number of water bodies and also as per its water area, Malda comes first in West Bengal. Maximum number of big and small water bodies also present in Malda. The wetlands census data of Ministry of Environment and Forests, Government of India, mentioned that Malda has 11 big freshwater wetlands out of 23 big wetlands of West Bengal (Anon, 1991). Due to regular recurring flood, Malda's wetlands area is increasing after every flooding season. Flood mainly due to high water rise in the river Ganges responsible for creating many major and minor water bodies in the district. Historical and geological evidences shows that flood were a regular recurring phenomenon during the ancient period of Gour Bengal. It was also found that the ancient capital Gour displaced its establishment very frequently to avoid its destruction from regular flood by the changing course of the river Ganges (Sharma- 2001, 2003a, 2003b 2005a).

The continuous soil erosion and embankment cutting has, however lead to loss of important settlements, mango orchards and highly productive agricultural lands which are usually subject to multiple cropping. New islands at the middle of the river Ganges also emerged but this economical and fertile land losses are not easily compensated by growth of many low laying areas which are also susceptible to regular annual flooding. Annual flood not only responsible for havoc loss of human lives, livestock, properties and prime agricultural lands but affects lakhs of people, making them landless refugees. Once the land eroded, engulfed by the river Ganges it never comes back. People those who are losing their settlements and prime agricultural lands can not allow to be resettle on these newly formed islands or on sandbanks due to political reason.

If wetlands of Malda are utlise to retain, store or to pass excess floodwater then these can be effective tools for flood control and flood mitigatation in the district to minimise the impact of flood. These natural resources are also very much beneficial for the environmental, ecological, economic and sociological development of the district, can lead to sustainable development if utlise in a wise way.

Every year thousands of people affected from flood and converted into as Flood Affected Refugees (FAR). These people can be resettling and rehabilitated along these wetlands sites. Conservation and wise use of these wetlands can provide them some alternative livelihood. New approach of irrigation and agricultural schemes for them can be implemented along these wetlands sites. As the district attracts largest migratory water birds in every winter season in West Bengal can be developed as the major eco-tourism spots, which will provide extra livelihood for those flood, affected peoples. Cultivation of fishes, makhana, shola, jalsingra also can provide them an alternative livelihood.

## Acknowledgements

I am very much thankful to the Irrigation Department, Agricultural Department, Manikchak, Kaliachak and English Bazar block administration, Block Development Officers, Irrigation Engineers from the flood and riverbank erosion affected areas. I am also thankful to the thousands landless flood affected refugees of Malda district who are shared their property, economic and livestock losses to me. At last I am greatly indebted to the Centre for Ecological Engineering for supporting the project.

## REFERENCES

Anon. (1991): *Wetlands of India – A Directory*. Ministry of Environment and Forests. Government of India, New Delhi.

Anon. (1998): *Report on the Status of Environment in West Bengal*. Department of Environment, Government of West Bengal.

Anon. (2000): *Agricultural Annual Plan of Malda district*. Agricultural Department. Government of West Bengal.

Anon. (2001): *Malda District Flood Contingency Plan and Programme*. Government of West Bengal.

Anon. (2002): *Master Plan for Flood Management and Erosion Control, Malda district, West Bengal*. Government of West Bengal.

Anon. (2003): *Flood Provisional Assessment of Damage and Restoration Requirement, Malda district*. Government of West Bengal.

Anon. (2003b): *Ganga Riverbank Erosion: A Burning Issue* (in Bengali). CPI (M), Malda District Committee, Malda, West Bengal.

Anon. (2004): *A Study Report on Control of Left Bank Erosion of River Ganga between Rajmahal and Farakka*. Irrigation and Waterways Department, Kolkata. Government of West Bengal.

Anon. (2005). *Flood Damage Report under Malda Irrigation Division in the district of Malda, West Bengal.* Irrigation and Water ways Department, Government of West Bengal.

Ahmed, I (1999): *Living with Floods – An Exercise in Alternatives*. The University Press Limited. Bangladesh.

Basu, S.K (1982): *A Geotechnical Assessment of the Farakka Barrage Project, Murshidabad and Malda district, West Bengal.* Geological Survey of India. Government of India.

Banerjee, S.N and Chakraborty, P (1983): Some observations on recent trends of shifting of the Ganga between Rajmahal and Ahiraon. *Journal of the Geological Society of India*. Vol: 24.

Centre for Ecological Engineering (2004): *River, Flood and Environmental Changes – An Analysis of Malda district, West Bengal.*

Chakrabarti, P ( 1998): Changing course of Ganga, Ganga-Padma River System, West Bengal, India – RS Data Usuase in USER

Orientation. *River Behaviour and Control*, Vol: 25, 1998. Journal of River Research Institute, West Bengal.

Dutta, B.N (2004): *Estimating and Costing in Civil Engineering*. UBSPD, New Delhi.

Garg, S.K (2001): *Irrigation Engineering and Hydraulic Structures*. Khanna Publisher, Delhi.

Ghosh, G and Bandhyopadhyay (1986): *Ganga – Bhagirathi Interfluves*. Contribution in Environmental Geology. Geological Survey of India. Government of India.

Ghosh, P (2004). *History of Malda District*. ( in Bengali ). Malda, West Bengal.

Khanna, P.K (2003): *Indian Practical Civil Engineers Handbook*. Engineers Publisher, New Delhi.

Lambourn, G.E (1918): *District Gazetteers of Malda*. The Bengal Secretariat Book Depot, Calcutta.

Mazumdar, S.K (2000): *Role of Farakka Barrage on the Disastrous Flood at Malda, West Bengal in 1998*. 8th ICID International Drainage Workshop, 31st January – 4th February 2000, New Delhi.

Mitra. J (2004): *Boundary Dipute between West Bengal and Jharkhand*. CPI (M) Malda District Committee, Malda, West Bengal.

Parua, P.K (1999): Erosion Problems of the river Ganga in the districts of Malda and Murshidabad in West Bengal. *Journal of American Society of Civil Engineers – India Section*, Vol: 13, No: 2, 1999.

Parua, P.K (2002): Fluvial Geomorphology of the River Ganga around Farakka. *Journal of Institution of Engineers*. Vol: 82, Kolkata.

Parua, P.K (2003): *Flood Management in Ganga- Brahmaputra – Meghna Basin: Some Aspects of Regional Co-Operation.*

Pawar, M.K (2006): River Morphology. *Newsletter for the Indian Society for Hydraulics*. Vol: 15, No: 1, June 2006. Pune.

Ray, S.C (2000): Erosional Problems in the Upstream of Farakka (in Bengali). Sechpatra, January-March - 2000. *Journal of Irrigation and Waterways. Government of West Bengal*. Kolkata.

Rudra, K (1993): Encroaching Ganga in West Bengal. *Indian Journal of Landscape System and Ecological Studies*, Vol: 16, No: 2. Calcutta University, Kolkata.

Rudra, K (1996): *The Farakka Barrage: An Interruption to the Fluvial Regime. Indian Journal of Landscape System and Ecological Studies*, Vol; 19, No: 2. Calcutta University, Kolkata.

Rudra, K (2000a): Living on the Edge: The Experience along the bank of Ganga in Malda district, West Bengal. *Indian Journal of Geography and Environment*, Vol: 5. Vidyasagar University, West Bengal.

Rudra, K (2000b): *Will the Ganga Outflank the Farakka Barrage?* (in Bengali). Desh Patrika, 9th December 2000. ABP Publication, Kolkata.

Rudra, K (2002): *The Saga of Bank Erosion: Malda and Murshidabad* (in Bengali). Mrittika, Kolkata.

Rudra, K (2004): *The Encroaching Ganga and Social Conflicts: The Case of West Bengal.*

Sen, P (2003): *Ganga Erosion at Malda* (in Bengali). Gour Malda Sambad, Malda.

Sengupta, J.C (1969): *Malda district Gazetteers*. Government of West Bengal.

Singh, S (1998): *Geomorphology*. Prayag Pustak Bhawan, Allahabad.

Singh, P *et al.* (1980): *Report of the Ganga Erosion Committee*, Government of West Bengal.

Sharma, A (1998): *An Ecological Concept of Flood Control in Malda district, West Bengal*, India. International Conference on Ecological Engineering, Science City, 23-27th November 1998. Kolkata. Kalyani University.

Sharma (2000): *Flood, Riverbank Erosion and Resource Depletion – A Case Study of Malda district, West Bengal.* Centre for Ecological Engineering, Malda, West Bengal.

Sharma, A (2001): *Problems of Landless Flood Affected Refugees of Malda, West Bengal.*

Sharma, A. (2002): *Riverbank Erosion in Malda – A Human Rights Perspective*. Malda, West Bengal.

Sharma, A (2003a): *Preserve or Perish. The Statesman*, 31st January 2003.

Sharma, A (2003b): *Conservation and Management of Wetlands for Flood Control and Sustainable Development – A Perspective from Malda district, West Bengal*. National Seminar on Conservation and Management of Natural Resources to Check Environmental Degradation, 11-12th January 2003. Malda College, Malda, West Bengal.

Sharma, A (2003c): *Wise use of Wetlands for Flood Control and Sustainable Development in Malda district, West Bengal.*

International Seminar on River, Floods: A Socio-Technical Approach, 14-16th February 2003. Visva-Bharati University, Santiniketan, West Bengal.

Sharma, A (2003d): Of Flood and Fragile Initiatives. *The Statesman*, 29th August 2003.

Sharma, A (2004): *Points to Remember*. The Statesman, 18th June 2004. Sharma. A (2006): *Ganga River bank Erosion – An In depth Analysis Report*. Centre for Ecological Engineering, Malda, West Bengal.

Sharma, A (2005a): Analysis of Flood and Ganga Riverbank Erosion at Malda. (in Bengali). *Bhugolika*, Vol: 11, No; 1and2, March 2005. Journal of Paschimbanga Bhugol Moncho, Kolkata.

Sharma, A (2005b): *Causes and Consequences of Flood and Riverbank Erosion in Malda district, West Bengal.* 15th Convocation of Diploma Engineers Association, 9-11th June 2005, Malda, West Bengal.

Sharma, A (2006a): Monsoon Menace. *The Statesman*. 14th July 2006.

Sharma, A (2006b): *Ganga Riverbank Erosion – An In-depth Analysis Report of Malda District, West Bengal.* Centre for Ecological Engineering, Malda, West Bengal.

Subramanya, K (1998): *Flow in Open Channels*. Tata McGraw-Hill Publishing Company Limited, New Delhi.

Sustain (1998): *The Green Book of Eco-Technology*. Austria.

UNDP (2004): *Environmental Issues*: West Bengal Human Development Report.

Vendre, V.M (2002): *Technical Report on Erosion on the Left Bank of the Ganga Upstream of Farakka in Malda district, West Bengal.*

# 4

# Land Sutability Analysis for Crop Agriculture and Shrimp Culture: Study on a Coastal Area of Bangladesh

*Bibi Khadiza, Mohammed Shariful Islam, Dr. Khan Rubayet Rahaman*

## ABSTRACT

The coastal zones of Bangladesh are quite famous for its high yield in shrimp culture. The local farmers cultivate mostly shrimp without assessing suitability of land for shrimp culture. At a result, the unsuitable land for shrimp is used for shrimp culture yielding low productivity on the one hand, and hindering crop culture for suitable land on the other. This paper addresses issues of suitability for crop culture and shrimp culture. Assessing the suitability of land for crop and shrimp culture requires a considerable effort in terms of information collection that presents both opportunities and limitations to decision makers. A comparative study has been carried out for selection of suitable sites for shrimp and crop culture using Geographic Information System (GIS). In this study, the land suitability for crop and shrimp cultural is assessed based on soil properties and water resource availability. It is then compared with the land use map to consider the conformity to it. Suitability of land for crop and shrimp requires separate assessment procedure of water availability, landform, soil texture and salinity of soil. Finally the suitable land for crop culture and shrimp culture has been identified. This paper concludes suggesting addition addition/subtraction of land for crop/ shrimp culture for optimum yield from the landuse.

**Keywords**: *Land suitability, crop culture, shrimp culture, coastal area, Bangladesh, GIS*

## Introduction

In the coastal areas of Bangladesh most of the lands are presently used for shrimp culture. But few years ago those lands were used for crop culture. As salinity intrusion to the coastal areas increases, crop culture is converted into shrimp culture. For this reason output from crop culture is decreasing. At the same time actual benefit from shrimp culture is not being yielded. Local farmers do not know the land suitability that means, which land is suitable for crop culture and/or shrimp culture. If proper use of land is not carried out, it fall effects on the profitability from land. At present, shrimp culture plays an important role in total GDP of the country (BBS, 1991). So land suitability is essential to identify the suitable land for shrimp culture along with crop agriculture to find out better production from land. Suitability analysis reduces the environmental and social impact of landuse on people (Dayawansa, 1991). It also helps farmers to select crops for cultivation.

Land suitability is identified as 'suitable' and 'not suitable'. A bit detail has been brought out classifying suitable lands into 'highly suitable land', 'moderately suitable land' and 'marginally suitable land'. Land suitability has been evaluated using some selected criteria which are based on soil quality, water quality, neighboring physical features and existing land use. The quality for suitability has been assessed based on studies of Bangladesh Agricultural Research Institute (BARI), Soil Research and Development Institute of Bangladesh (SRDIB), Bangladesh Water Development Board (BWDB) and Local Government Engineering Department (LGED) of Bangladesh. Land suitability is determined under GIS environment operationalizing multi-criteria analysis (ESRI, 2006). The map features used for such analysis are, settlement map, road map, soil $P^H$ map, soil salinity map, soil type map, electricity coverage map, irrigation coverage map etc (Salam and Ross, 2000; Yamamoto and Sukchan, 2002)

This paper addresses delineation of existing areas for crop culture and shrimp culture and classification of the suitable areas for crop culture and shrimp culture. For this study Chandpur area of Ghatbhog union under Rupsha Upazila in Khulna district is selected. This area is a salinity affected area and it is situated in the coastal zones of Bangladesh. In this area two types of agricultural land use are found: one, crop culture and another shrimp culture. Though few cases identified for crop agriculture most of the land parcels are being used for shrimp culture without judging whether those are suitable for shrimp culture.

## Land Suitability

Land suitability may be defined as "the fitness of a given type of land for a specified kind of land use" (*Food and Agriculture Organization of the United Nations, 1985,* p.-61). The suitability is the aptitude of a given type of land to support a defined use. The process of land suitability classification is the appraisal and grouping of specific areas of land in terms of their suitability for a defined use. 'Suitable lands' are meant for lands on which sustained use is expected to yield benefits which justify the inputs, without unacceptable risk of damage to land resources, on the other hand, 'not suitable' lands are defined for the lands having qualities appearing to preclude sustainability for the considered land use. In this paper 'suitable' and 'not suitable' are symbolized as S and NS. Suitability is further classified into three subgroups, namely, highly suitable, moderately suitable and marginally suitable. Highly suitable land means land which has no significant or only minor limitations to the sustained application of the given land utilization. Moderately suitable land means land which has limitations that are moderately severe for sustained application of the given land utilization. The limitations will reduce productivity or benefits and will increase the required inputs. Marginally suitable land means land which has severe limitations for sustained application of the land utilization. In this study 'highly suitable', 'moderately suitable' and 'marginally suitable' lands are symbolized by $S_1$, $S_2$ and $S_3$.

## Criteria for Suitability

Land suitability of shrimp and crop culture is identified based on some selected parameters which are, soil characteristics (soil type, soil pH and soil salinity), salinity of soil, elevation of land, location/connectivity of neighboring physical features (roads, electricity, settlement, farm and market), and water quality (water $P^H$ and water salinity).

Criteria of suitability have been collected from different relevant research institutes for the study area which is also applicable for all other coastal areas of Bangladesh. The suitability criteria for different type of suitability have been depicted in Table 4.1 and Table 4.2.

**Table 4.1. Suitability criteria for crop culture**

| *Parameter* | *Highly suitable* | *Moderately suitable* | *Marginally suitable* |
|---|---|---|---|
| Land elevation | High | High | Medium |
| Soil pH | 6.6-7.3 | 5.6-6.5 | 4.5-5.5 |
| Soil salinity | Less than 2 EC (ML mos/cm) | 2-4 EC (ML mos/cm | 4-6 EC (ML mos / cm |
| Water salinity | .55-.75 EC (ML mos/cm | .35-.55 EC (ML mos/cm | .15-.35 EC (ML mos / cm |
| Water pH | Above 9 or less than 6 | 4.5 - 6 | 3.5-4.5 |
| Soil rigidity | Hard | Medium hard | Medium hard |
| Soil Roughness | Plain | Slightly plain | Slightly plain |
| Soil type | Irsordi and Gopalpur | Bajoa | Satla |
| Irrigation Coverage | Should be exist | Should be exist | Should be exist |

*Source:* BARI, SRDIB, BWDB and LGED, 2006

**Table 4.2. Suitability criteria for shrimp culture**

| *Parameter* | *Highly suitable* | *Moderately suitable* | *Marginally suitable* |
|---|---|---|---|
| Land elevation | Low | Low | Medium |
| Soil pH | 6.5-9 | 5.5-6.5 | 4.5-5.5 |
| Soil salinity | Above 8EC (ML mos / cm) | 5-8 EC (ML mos / cm | 4-5 EC (ML mos / cm |

| *Parameter* | *Highly suitable* | *Moderately suitable* | *Marginally suitable* |
|---|---|---|---|
| Water salinity | Above .75 EC (ML mos / cm | .75 EC (ML mos / cm | .65 EC (ML mos / cm |
| Water pH | 6.5-8.5 | 5.5-6.5 | 4.5-5.5 |
| Soil type | Barisal | Harta | Satla |
| Distance from river | 1000 m | 1500 m | 2000 m |
| Soil rigidity | Soft | Soft | Medium soft |

*Source:* BARI, SRDIB, BWDB and LGED, 2006

## ANALYSIS AND FINDINGS

### Existing area used for Crop and Shrimp Culture

At present, the study area is identified for different type of landuses. Crop and shrimp land are identified as an agricultural use in the study area. Some land are found for crop culture, some are found for shrimp culture and some land are found which are used for both crop and shrimp culture. The amounts of different type of landuse in existing situation of the study area are quite interesting. 81.40 acre lands are used for crop culture which represents 27.59% of total land, and only 8.38 acre land are used for shrimp culture. Though such small amount of lands used for shrimp culture only, 66.82 acre land are used for both crop and shrimp culture simultaneously. Other types of landuse that are, settlement, water body, plantation, roads, river, cannel, etc. are comprised of 122.01 acres of land which represent 41.37 % of total land. Please see also Figure 4.1 and Figure 4.2.

### Suitability of Crop Culture and Shrimp Culture

Suitable area for crop culture is identified as; 69.79 acre land is highly suitable and 82.67 acre land is not suitable representing 24% and 28% of total land area in the study area respectively. Again, 10.9 acre land is moderately suitable and 9.6 acre land is marginally suitable; which are 3.69% and 3.25% respectively (Table 4.3 and Figure 4.3).

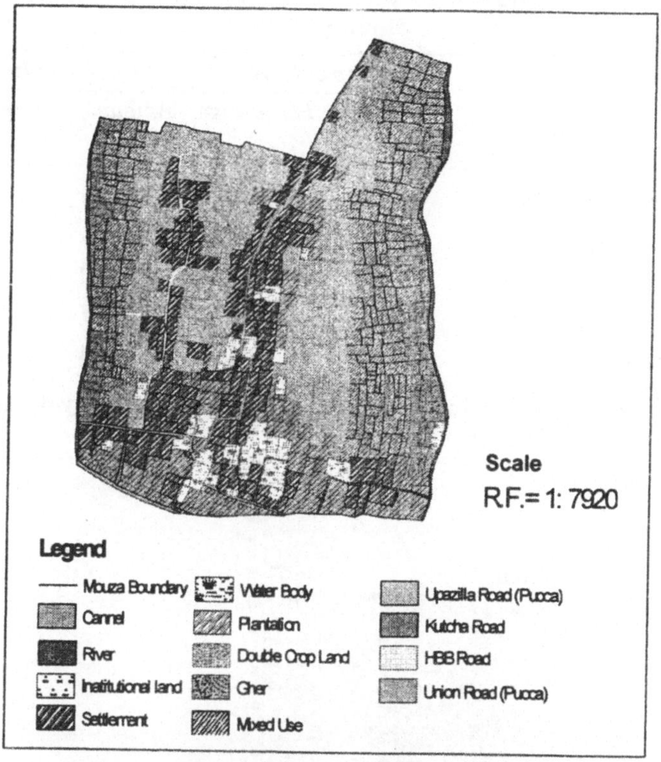


**Fig. 4.1: Existing landuse map of the study area**
*Source:* Land-use survey, 2006

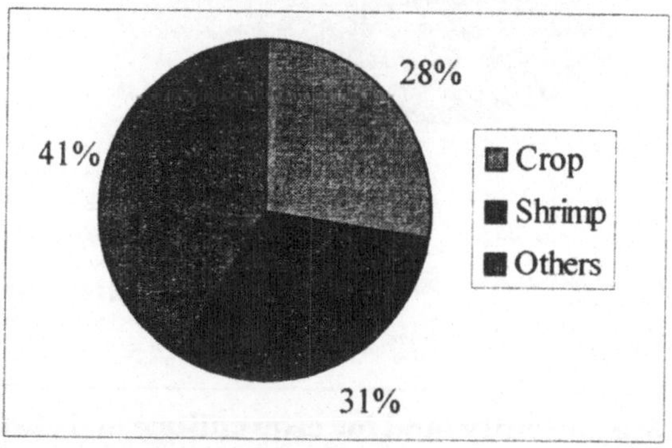


**Fig. 4.2: Percentile distribution of existing land-use**
*Source:* Land-use Survey, 2006

**Table 4.3. Suitable area for crop culture and Shrimp culture**

| *Category* | *Crop culture* | | *Shrimp culture* | |
|---|---|---|---|---|
| | *Amount (acre)* | *Percentage* | *Amount (acre)* | *Percentage* |
| Highly Suitable | 69.79 | 23.65 | 39.75 | 13.48 |
| Moderately suitable | 10.9 | 3.69 | 10.14 | 3.44 |
| Marginally suitable | 9.6 | 3.25 | 00 | 00 |
| Not suitable | 82.67 | 28.02 | 123.07 | 41.72 |
| Others | 122.04 | 41.36 | 122.04 | 41.36 |

*Source:* Data produced from Multi-criteria analysis in GIS environment, 2006.

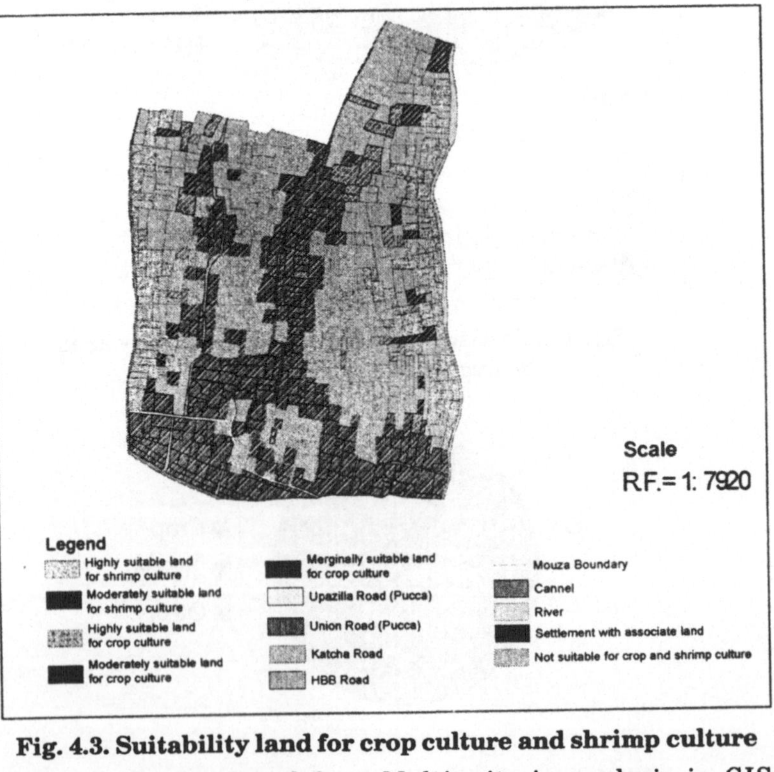


**Fig. 4.3. Suitability land for crop culture and shrimp culture**

***Source:*** Map produced from Multi-criteria analysis in GIS environment, 2006.

For shrimp culture, 39.75 acre land is highly suitable and 123.07 acre land is found which are not suitable which represent 13.48% and 41.72% of total land respectively. Table 4.3 also shows that 10.14 acre lands are moderately suitable for shrimp culture which represents 3.44% of total land. No land is found for marginally suitable shrimp land.

Findings in the concrete form can be drawn as, 30.57% land is suitable for crop culture, 16.67% land is suitable for shrimp culture and 11.58% land is not suitable for both crop culture and shrimp culture. 81.40 acre land is used for crop culture at present but from suitability analysis it is found that, 90.19 acre land is suitable for crop culture. So, 8.79 acre land can be added for crop cultivation (Table 4.4). Again, 91.56 acre land is used for shrimp culture at present but from suitability analysis it is found that 49.19 acre land is suitable for shrimp culture. So, 42.37 acre land should be transformed from shrimp culture to other use and/or crop culture.

**Table 4.4. Comparison between existing situation and suitable situation**

| *Category* | *Existing (Acre)* | *Percentage* | *Suitable (Acre)* | *Percentage* | *Remarks* |
|---|---|---|---|---|---|
| Crop land | 81.40 | 27.59 | 90.19 | 30.57 | + 8.79 |
| Shrimp land | 91.56 | 31.03 | 49.19 | 16.67 | -42.37 |

*Source:* Data produced from Multi-criteria analysis in GIS environment.

## SUMMARY AND CONCLUSION

Land suitability for crop and shrimp culture is important to yield high productivity from land. From this study it is found that 90.19 acre lands are suitable for crop culture but in present situation 81.40 acre land are used for crop culture. So it can be said that if 8.79 acre land are added for crop culture outcome of land.

From this study it is found that 8.79 acre land should include in crop culture for better utilization of land. One the other hand 24.45 acre land should be included in shrimp culture. From the analysis it is also found that 33.38 acre land are not suitable for crop or shrimp culture. Those unsuitable lands can be used for physical development like institutional establishment, government office establishment etc.

Land suitability analysis for crop and shrimp culture is important for better use of land. For better utilization of land it is essential for identification of suitable use of land (Perera and Tateishi, 1994.). For wrong utilization of land, it creates land use conflict in the south western part of Bangladesh. If suitable location for crop and shrimp culture can be chosen properly, it helps to remove land use conflict. If farmers use suitable area for correct purposes, it gives better profit and it also helps farmer to choose crop or shrimp culture at the time of cultivation. Proper utilization of land would also help reducing environmental vulnerabilities.

Development plan for an area might require landuse pattern and its optimum use. This study would help planners to plan for appropriate development and land utilization. If planners consider the suitable land for crop and shrimp culture it helps to optimum use of land. This study has scope for further stepping; landuse characteristics has been the base for the criteria selection for this study but climatic conditions and seasonal variations which are important aspects for agricultural yields, have scope through this study to be studied.

## REFERENCES

Bangladesh Bureau of Statistics (BBS), 1991. Population Census; community series-Khulna District, Bangladesh Bureau of Statistics, Dhaka.

Dayawansa, N.D.K., 2006. *Land Suitability Identification for a Production Forest through GIS Techniques*, Department of Agricultural Engineering, Faculty of Agriculture, University of Peradeniya, Sri Lanka.

ESRI, http://www.esri.com/industries/agriculture/research/suitability.html, explored on October 25, 2006.

Food and Agriculture Organization (FAO), 1978. Report on the Agro-ecological Zone Project, Vol.1, Methodology and Results for Africa, World Soil Resource Report No 48, FAO, Rome.

Hossain, M. S. and Lin C. K., 1998. "Land Suitability Analysis for Integrated Coastal Zone Management in Cox's Bazar, Bangladesh:

A Remote Sensing And GIS Approach, Integrated Tropical Coastal Zone Management", Asian Institute of Technology, Thailand, 2003.

Perera, L.K. and Tateishi, R., 1994. "Do remote sensing and GIS have a practical applicability in developing countries?" *Int. Journal of Remote Sensing*, Taylor and Francis, Vol. 16, 1, p 18-35.

SRDI, 1989. *Khulna, Bangladesh*, Soil Resource Development Institute.

Salam, M. A. and Ross, L. G., 2000. "Optimizing sites selection for development of shrimp (Penaeus monodon) and mud crab (Scylla serrata) culture in Southwestern Bangladesh" Department of Aquaculture, Faculty of Fisheries, Bangladesh Agricultural University, Mymensingh- Bangladesh.

Yamamoto, Y. and Sukchan, S., 2002. "Land Suitability Analysis Concerning Water Resource and Soil Property" *Japan International Research Center For Agricultural Sciences (JIRCAS)*, Japan.

# 5

# Sustainable Livelihoods: An Approach and a Framework for Analysis

*Chandima Daskon*

*Together you and I will begin to build the new society,*
*... a society in which each of us has the chance*
*... to grow, to achieve, to contribute,*
*... to create dignity for ourselves, and*
*... not for ourselves alone, but*
*... for others also; a society in which*
*... each of us has a stake, a share and*
*... we will give back to our children what they deserve –*
*... a heritage of hope*

—White Paper for International Development 1997

## Introduction

After the structural perspective and neo-Marxism of the 1970s and 1980s, a more productive actor-oriented perspective was adopted in the development process. Like the predecessors, it has also emphasized inequalities in distribution of assets and power, but importantly it recognizes that people make their own history and it even opposes the views that economic concerns are necessarily of primary importance. The Sustainable Livelihood Approach (SLA) emerged from this concern with a holistic view of tackling poverty through putting poorer people and their priorities at the centre of development process. The Sustainable Livelihoods Approach, the framework and principles are now employing as tools to guide, understand and practice the planning and implementation of development activities. Modern livelihood studies found their intellectual inspiration in the general understanding of the lives of poor people advocated by Chambers and Conway (1992).

The Chambers' notion 'putting the last first' paradigm has given strong impetus for the emergence and establishment of the SL approach in development context. Today, livelihood perspective has increasingly adopted as a concept for better understanding poverty and socio-economic inequalities in the world. Especially the approach is now becoming more popular among the development geographers as an effective way to addressing human poverty and well-being (e.g. de Hann and Zoomers (2005) and Bebbington (1999). Since the 1990s, the livelihood perspective has gained significant momentum as a way of looking at development by putting people at the centre, stressing their active role in exploring their own opportunities and coping with changes. The livelihood approach has become opposed to earlier development theories and models that tended to portray people as passive victims of top-down decisions and the Eurocentric development. Therefore, the livelihood approach goes beyond the material aspect of life recognizing multidimensional aspects covering not only economic, but also political, social, cultural and ecological aspects of life of people.

The purpose of this chapter is to provide a comprehensive understanding about the Sustainable Livelihood Approach (SLA). In this regard this chapter is structured as follows.

The first section discusses about the theoretical path towards the emergence of the SL approach. It draws a broad theoretical background in order to establish this approach within the current development debate. The section briefly elaborates development approaches under classical, historical, dependency and alternative themes. Secondly the chapter explains the conceptual and organizational routes of the current SL approach. Based on a broader literature review, the third section of this chapter attempts to give a holistic understanding of the term 'livelihoods' with different interpretations. The fourth section describes the guiding principles of the SL approach which are the key considerations of the current livelihood interventions. The fifth section gives a substantial description about the SL framework with its key components as recognized under *vulnerability context, livelihood assets, transformation structures and processes* and *livelihood strategies*. The sixth section devotes to discuss about the practical implications of the SL approach with a number of case study examples and the projects recently undertaken by the different development agencies. The final section draws the strengths and weaknesses of this approach based on the criticisms and problems encountered by development agencies and researchers during researching, planning, implementing and evaluating livelihood projects.

## Towards the Sustainable Livelihood Approach (SLA)

The development theories and approaches emerged over last five decades provide logical prepositions of the changes of political, economic, social, ethnic, cultural, and ecological dimensions of development thinking. However, it is important to notice, though the classical development theories are superficially neat to characterize as modernization, dependency and alternative approaches, neither rural development theories nor processes have indeed undergone in such an uncluttered manner. Nevertheless, these paradigms have significantly influenced to founding the SL approach through prioritizing people's genuine capabilities and strengths of living.

This section examines the changing theoretical debates of development towards the establishment of the SL approach. As many scholars assert (e.g. Escobar, 1992; Schech and Haggis, 2000) reviewing development as a 'discourse' is straightforward in getting different meanings attached to the term development that has been produced within and by a particular context. In order to do that, development theories are defined here under four major approaches following Potter *et al.* (1999, 2004)' classification. They categorize development theories and approaches under classical, historical, dependency and alternative approaches (see Fig. 5.1). This explains the logical propositions, which purport to explain how development has occurred in the past, and/or should occur in the future.

## Classical/Traditional Theories of Development

The Classical development theories are mainly led by Adam Smith and David Ricardo and both has equated economic development with 'growth of world trade and the low of comparative advantage'. The whole idea about the classical approaches represents 'unbalanced economic growth', and so often these theories highlight a dualistic structure of social organization as a result of the inequalities created by the growth process. These dual structures are often described as top-down, backward and advanced, barbarian and civilized, and traditional

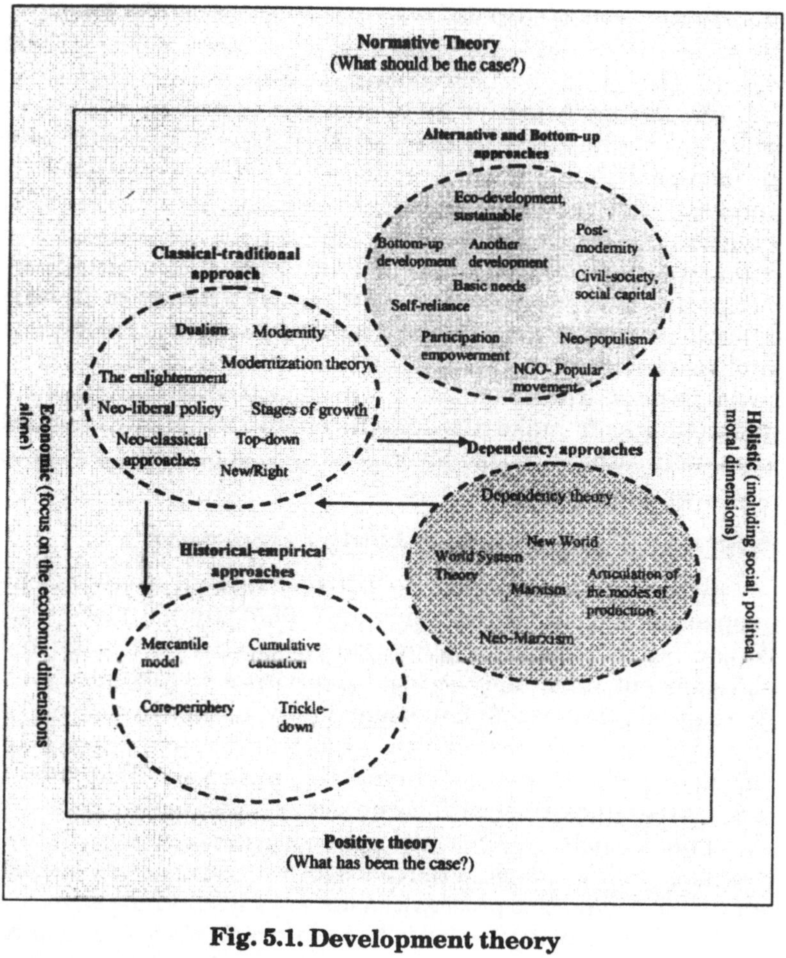


**Fig. 5.1. Development theory**

*Source:* (Potter *et al.*, 2004)

and modern. Potter *et al.*, (2004), Stohr, (1981) and Stohr and Taylor, (1981) collectively named these classical theories as 'top-down' approaches. Arthur Lewis, one of the major 'gurus' of this classic era stresses that, *our subject matter is 'growth' and not 'distribution* (Esteva and Prakash, 1998)[1]. Quoting Nehru, Gunnar Myrdal, (1968) notices the characters of this classical development theories . . . in order to be 'developed', we have to get out of these traditional ways of thinking, traditional ways of acting, traditional ways of production, traditional ways of distribution and traditional ways of consumption. We have got to get out of all that into what might be called more modern ways of doing so . . ." Hirschman, (1958), asserts that even though, the polarization is an inevitable and spontaneous process of the early stages of this type of growth and development, the core will lead to the 'trickle-down' of growth-inducing tendencies to backward regions. In such a case, according to Hirschman, governments should not intervene to reduce inequalities, for at some juncture in the future the search for profits will promote the spontaneous spin-off of growth inducing industries to backward regions. According to Friedmann and Weaver, (1979), *inequality is efficient for growth, but equality is inefficient*. However, the main objective of the classical theorists was to make a 'wholesale change in underdeveloped societies in order to brake the vicious cycle of poverty, ignorance and low productivity'.

## Historical approaches to development

Bringing more critics to above classical theories the pioneering scholars like Friedmann, (1966), Myrdal, (1957) and Vance, (1970) provide an empirical perspective on change and development, under the *historical approaches*. As Myrdal argues, the capitalistic development is inevitably marked by deepening regional and personal income and welfare inequalities by developing the theory of 'cumulative causation'.[2] This theory argues that once differential growth has occurred, internal and external economies of scale will perpetuate the patterns. However, Myrdal employs the concepts of 'backwash' and 'spread' effects to explain this process and his work leads to the advocacy of strong state policy in order to counteract what was seen as the 'normal tendency of the capitalist system' to foster 'increase regional inequalities'. Developing the 'core-periphery' model

Friedmann (1966), argues that, the economic development will ultimately lead to the convergence of regional income and welfare differentials. Core regions have recognized as the major centres of innovative changes, while all other territories consists of the 'peripheral' regions. They are dependent on the core regions and their development is largely determined by institutional behaviours of the core. Vance (1970) provides a historical and evolutionary viewpoint of the national patterns of development and changes through developing the mercantile model.

## Dependency Theory and development approaches

The dependency theory was developed as a critique of and countered to above two approaches. This was the first major Third World challenge to Europe-centred academic discourse and according to Tucker, (1999) it provides more logical explanations to the process of underdevelopment in the Third World. The approach has given much concern on indigenization of development thinking based on the conditions that are encountered in the Third World. Leading works of the Latin American scholars', like Prebisch (1950), Furtado (1964, 1965, 1969), Dos Santos (1970, 1977), Cardoso (1976), Paul Baran (1973), Gundar Frank (1966, 1967) and Wallerstein (1974, 1980) encapsulate the phrase of 'development of underdevelopment' emphasizing the chain of dependent relations and its expansion to describe the process of underdevelopment. As the approach argues, the dominant capitalist powers encourage the transformation of their political and economic structures to poor nations in order to serve capitalist's interests. Development process has conceived largely in terms of economic growth, industrialization and liberal democracy. While criticizing modernization theorists this did not question the desirability of development. Neither ecological implications of growth model nor rights of tribal peoples were considered. Within this approach poverty was irrelevant although assets and levels of consumptions were measured in order to indicate how far specific social classes were disadvantaged. However, the approach's interest was to asses the 'accumulation potential' rather than any kind of equal distribution. As far as economic consideration dependency theorists were no different from classical and historical development theorists.

However, according to Ellis and Biggs, (2001) the first 'paradigm shift' to rural development has occurred in early 1960s. The 'small-farm efficiency' paradigm (e.g. Schultz, 1964) was considered as the engine of 'growth' and development. According to this paradigm, the traditional or the subsistence sector possesses a negligible prospect for growth and therefore it could only play a passive role in economic development (Jhonston and Mellor, 1961; Lewis, 1954). As the paradigm explains the subsistence sector supplies resources to modern sector which, envisages as containing large scale 'modern' agriculture in addition to the manufacturing industry. Then the notion of livelihood has become more or less an economic concept and their changes were assumed to follow a straight line towards higher standards of living. As de Hann and Zoomers, (2005) point out, livelihood even as a concept has almost completely vanished in development geography due to the dominant structural perspectives of dependency and neo-Marxist approaches in development geography. As agricultural economists led the economic analyses the attention for differentials of resource accessibility and people's skills were very considerably limited. Household as an economic unit has modelled by choosing the income opportunities that provided the biggest pay-off. Therefore, the 'human progress' and the quality of development have often represented by the statistical figures which, move up and down in the charts of 'progress'.

## *What now?* Another development

Diversity and decentredness must be placed at the top of the agenda in reformulating the question of development. The cultural question, the question of the meaning of development, is a central one. For one society to claim universal desirability, while turning its back on others from whom it is convinced it has nothing to learn, is nor only cultural elitism but cultural racism

(Tucker 1997, p. 13)

As a result of the prolonged critics and the failures of 'growth' concept new theoretical debate was forming during the 1970s. These criticisms have awakened a new interest about 'people' and the development process was re-adjusted as more appropriate to local realities and especially as workable for the poor people's desires. Many authors have began to unfolding grassroots going beyond the premises and promises of modernity and growth, re-inventing and creating afresh new intellectual and institutional framework for the people at the grassroots. Despite some big gains especially in health and education sectors the scale and awfulness of deprivation among rural people of the developing world, remained an outrage. Therefore, the question remained whether these new thoughts and initiatives can succeed in those domains where the old paradigms have failed.

The major concern on was appeared through the Seventh Special Session of the United Nations General Assembly and the allied publication by the Dag Hammarskjold Foundation *What Now?*. This whole idea was synthesized as an 'alternative approach' that uses as an umbrella concept to include a broad sweep of changes in thinking regarding development and how best to achieve it from the late 1970s (Elliott, 2006 p20). Stohr and Taylor (1981) refer this as '*development from below*' and the ultimate objective is 'the full development of a region's natural resources and human skills'. Chambers (1983) refers this as 'putting the last first' paradigm by reversing the power relations among development actors and local people. Chambers intentionally introduces this whole idea as a 'new paradigm' to mean a 'coherent and mutually supporting pattern of concepts, values, methods and action, amenable to wide application, (1995, p. 2).

Even though the term 'alternative' is an uninformative expression, this stresses the need for self-reliance to be seen as central to the development process and emphasis should be placed on 'internal' rather than 'external' (Potter *et al.*, 2004). It has also increasingly suggested that development should focus on the *basic needs* of people while ecologically and culturally sensitive with a greater extent of public participation (Chambers, 1995a; Elliott, 2006; Potter *et al.*, 2004). According to Schech and Haggis, (2000, p 12), besides food, shelter, and clothing popular participation in the decision-making that affect the lives and livelihoods of the people, and individual freedoms also included as basic needs of people. More interestingly as Chambers asserts this new paradigm reverses the traditional core-periphery relationship;

> *"Normal professionals face the core and turn their backs upon the poor, New ones by standing on their head face the periphery instead . . ."* (Chambers 1995 p 14)

According to de Hann and Zoomers, (2005) this new actor-oriented perspective interests in the world of lived experiences, the micro-world of family, network and community. This draws attention to related issues such as poverty, vulnerability and marginalization and micro-orientation has became prominent that accompanied by a clear focus on local actors, often households. Within this paradigm, household strategies are identified as the modern vogue in order to capture the behaviour of poor people, while recognizing its importance for bridging the gap between micro-economics. For example, as Schmink, (1984) assert that *in response to the opportunities and constraints defined by broad historical and structural processes, the domestic unit is conceived of as mediating a varied set of behaviours that are themselves conditioned by the particular makeup of those most basic economic entity.* However, Guyer and Peters, (1987)'s judgment of household strategies bring critical conclusion which is equally valid for the Sustainable Livelihood Approach as well. *'The major shortcoming of the structural-functional and economic approaches to the household is the neglect of the role of ideology. The socially specific units that approximate 'households' are best typified not merely cluster of task oriented activities, not merely as a place to live/eat/work/reproduce, but as a source of identity and social markers. They are located in structures of cultural meaning and differential power* (1987, p 209). This background has given momentum to emergence of the SL approach as discuss in the following section.

## Conceptual and Organizational routs of the Livelihood approach

According to Ellis and Biggs, the idea about Sustainable Livelihood Approach has emerged during the 1980s and 1990s as a second paradigm shift in entire rural development process.

However, the origin of the concept of livelihood and the approach is seemed overlapping with many development theories, approaches and debates. As Ellis and Biggs further emphasis, most of the development ideas and concepts that first appear in one decade often gain strength in the following decade and only begin to affect practically in a widespread way ten or fifteen years after they were first put forward. This is a clearly a true factor about the SL approach; because, even though the SL approach is increasingly using in the current decade, the idea about livelihoods and sustainability developed through the 1980s and 1990s (Chambers, 1983; Chambers and Conway, 1992) and from famine analysis of the 1980s (Sen, 1981; Swift, 1989).

The livelihood approach represents a composite of ideas and interests in which a number of strands in development debate have come together. In a perfunctory glance, the term livelihood is synonymous and sometimes overlapping with the terms like employment and job. But, the term livelihood captures more complex and diverse strategies of living, than what is meant by employment (Arce, 2003). It has appeared to subsume both modern sector employment and other kinds of marginal, parallel, part time, informal or new forms of economic or non-economic activity and significantly it introduces the social dimension of sustainability and equity (Singh and Wanmali, 1998). However, in a simple term, livelihood comprises a number of activities engage that neither require any formal agreement nor are limited to a particular trade. It may or may not involve money and is self directing. Sometimes livelihoods are may based on the incomes derive from jobs but also on incomes derive from assets and entitlements as well[3].

This idea was established as a concept with growing legitimization through several major international forums. The Brundtland commission in 1987 for example, introduced the concept of *Sustainable Livelihood referring to resource* ownership and access, basic needs and livelihood security especially in rural areas (WCED, 1987). In 1992, United Nations Conference on Environment and Development (UNCED), in Agenda 21 expanded this concept by advocating the achievements of '*Sustainable Livelihoods for all*'. This has mainly focused on the issues of unsustainable production and consumption as relevant for both the North and the South. The major emphasis has given to integrative power of the sustainable livelihood concept offering ways of linking socio-economic and ecological policy considerations into a cohesive policy-relevant structure. The Copenhagen Declaration calls for 'Secure Livelihoods for all' in their third commitment: '*we commit ourselves to promoting the goal of full employment as a basic priority of our economic and social policies, and to enabling all men and women to attain secure and sustainable livelihoods through freely chosen productive employment and work*' (United Nations, 2000).

The Beijing Conference has further expanded the livelihood concept emphasizing the importance of enhancing livelihood strategies of women (United Nations, 1995). According to deHann, (2000), Geography becomes the first discipline that livelihood concept is academically established and currently uses in a wide spectrum. As de Hann (2000) identifies the notion of *genre de vie* in French geography was the first conception of livelihood in modern geography. Although the concept was completely vanished during the structural analysis and neo-Marxist approaches, the concept was re-emerged with the actor, agency and place based development analysis.

## Definitions of 'Livelihoods'- A Holistic Understanding

It is clear that the term livelihood is a highly complex. According to Carney, (1998b) the current definition of SL is normative. As Wanmali, (1999) notices, this concept still remains difficult to operationally define outside specific, subjective and situational contexts. Even though, the proliferating literature on sustainable livelihood provides a number of definitions and explanations about 'livelihood', according to Carswell et al (1997) the '*definitions of Sustainable Livelihoods are often unclear, inconsistency and relatively narrow. Without clarification, there is a risk of simply adding to a conceptual muddle. . . .*'

WCED, (1987) for example, provides a detailed explanation based on the concept of Sustainable Livelihood Security' (SLS). In this context, livelihood refers to *adequate stocks and flows of food and cash to meet basic needs..a household may be enable to gain sustainable livelihood security in many ways-through ownership of land, livestock or trees; rights to grazing, fishing, hunting or gathering; through stable employment with adequate remuneration; or through varied repertoires of activities* ( 1987 p 2-5).

According to the Stockholm Environmental Institution (SEI) livelihood is *the creation of conditions that are (self-)supportive of sustainable development in human, natural and economic systems, which, whilst safeguarding resources and opportunities for future generations, provides individuals with means to provide themselves with food, shelter and an acceptable quality of life. ...*[4]. The International Institute for Sustainable Development (IISD)[5] denotes Sustainable Livelihoods as being, *concerned with people's capacities to generate and maintain their means of living, enhance their well-being, and that of future generations.* According to Oxfam (GB), Sustainable Livelihoods are those that *allow people to cope with and recover from shocks, maintain quality of life over time, and provide the same or better opportunities for all, now and in the future.*[6] With the special focus on the developing world UNDP asserts livelihoods as the *means, activities and entitlements by which people make a living*[7]. Therefore, livelihood system is a dynamic realm that integrates both the opportunities and assets available to people for achieving their goals and aspirations as well as interactions with and exposure to a range of beneficial or harmful ecological, social, economic and political perturbations that may help or hinder community's capacities to make a living. Wallmann, (1984) articulates livelihoods in a very descriptive way:

> Livelihood is never just a matter of searching shelter, money and food. It is equally a matter of ownership and circulation of information, the management of skills and relationships and affirmation of personal significance ..and group identity. The tasks of meeting obligations, of security, identity and status are crucial to livelihood as bread and shelter (p 32).

But, this is not to say that livelihood is not a matter of material well-being; but rather that it includes non-material aspect of well-being as well. Bebbington (1999) for example, provides a holistic definition of livelihood;

> A livelihood encompasses income, both cash and in kind, as well as the social institutions (kin, family, and village), gender relations, and property rights required to support and to sustain a given standard of living. A livelihood also includes access to and the benefits derived from social and public services provided by the state such as education, health services, roads, water supplies and so on ( 1999 p 2022).

Therefore, the understanding of sustainable livelihoods is holistic and meaningful, when it meets social, economic, cultural and spiritual needs of all members of a community including human, non-human, present and future-and safeguards of cultural and biological diversity. As Scoones, (1998) notices however, the key issue need to be included both in livelihood definitions and analyses is;

> ..given a particular context (of policy setting, politics, history and socio-economic conditions), what combination of livelihood resources (different types of assets) result in the ability to follow what combination of livelihood strategies (agricultural intensification, livelihood diversification and migration) with what outcomes? Of particular interest of the SL framework are the institutional processes (embedded in a matrix of formal and informal institutions and organizations), which mediate the ability to carry out such strategies and achieve (or not) such outcomes (1998 p 3).

The seminal work of Robert Chambers and Gordon Conway, introduces Sustainable Livelihoods as a practical concept for the 21$^{st}$ century. They developed the working definition of livelihoods as

> ..the capabilities, assets (stores, resources, claims and access) and activities required for a means of living; a livelihood is sustainable when it can cope with and recover form stresses, and shocks and maintain or enhance its capabilities and assets, and provide sustainable livelihood opportunities for the next generation; and which contributes net benefits to other livelihoods at the local and global levels and in the short and long term (Chambers and Conway, 1992)

This has become the foundation for many of the above explanations and other conceptualizations adopted by the different development institutions[8]. The research team of Institute of Development Studies (IDS) that is led by Ian Scoones (1998) proposes a modified version of the Chambers and Conway's livelihood definition. According to Scoones, a livelihood comprises the capabilities, assets (including both material and social resources) and activities required for a means of living. A livelihood is sustainable when it can cope with and recover from stress and shocks maintain or enhance its capabilities and assets, while not undermining the natural resource base.

The main difference between these definitions is that it does not include the requirement that for livelihoods to be considered sustainable they should also *...contribute net benefits to other livelihoods'.*

Adopting Chambers and Conway's definition, Ellis (2000) expands the original definition of livelihood, considering the local *institutional context* that always affects in building sustainable livelihood strategies. As Ellis believes, local institutions work differently from those operating over a larger territory with overlaps and conflicts between them. Hence resource *accessibility* is identified as a crucial issue in describing sustainable livelihoods. In this sense Ellis defines that *a* 'livelihood comprises the assets, the activities and the "access" (mediated by institutions and social relations) that together determines the living gained by the individual or household; "accessibility" to livelihood resources and opportunities may change due to shifting local institutional context surrounding their livelihoods (2000:10).

However, there is an inconsistency in the current definitions of livelihood. As Chambers and Conway (1991) argue that, many livelihoods are largely predetermined by accident of birth. Livelihood activities are therefore 'culturally' determined and people may be born, socialized and apprenticed into an inherited livelihood as a cultivator, a pastoralist, a forest dweller, or as a fish-person. Therefore local cultural context is a crucial phenomenon in a holistic definition of livelihood. According to Toner, (2003) livelihoods are multi-sectoral, that all aspects of peoples' lives will impact on the livelihoods choices that they make, and that livelihoods are embedded within specific institutional context. Hence, the livelihood approach is naturally responsive to people's interpretations and their priorities of livelihood choices. These priorities and livelihood choices are fundamentally shaped by the local values and customs that have been practiced over generations.

## The Principles: A Guide for Livelihood Sustainability

The livelihood approach constitutes a set of principles that encapsulate the best development practices of numerous individuals and agencies. These principles have their foundations in the learning process approaches (Bond and Hulme, 1999) and can be applied to any type of development activity (Ashley and Carney, 1999). These principles are essentially a checklist of current best practice and they also reflect the concerns and assumptions that underpin the theoretical framework discuss in the following section. Carney (2000) distinguishes these principles as normative and operational (Box 5.1).

**Box 5.1. Livelihood guiding principles**

| | | | | |
|---|---|---|---|---|
| Normative | 1 | **People cantered**: (a) responding to people rather than resources or services (b) Involvement of wider range of beneficiaries | Operational | **Conducted in partnership**: formed with poor people as well as public and private sector |
| | 2 | **Participatory and responsive**: process to listen and respond to the poor | | **Multi-level and holistic**: (a) Primarily the micro informs the macro level. (b) Micro and meso levels are to help build on strengths at micro (c) study changes across a range of livelihood aspects |
| | 3 | **Dynamic and Sustainable**: (a) In all four dimensions-environmental, social, economic and institutional (b) Long term viability of interventions and sustainability | | **Disaggregated**: recognizing that livelihood profiles differ between groups and gender |
| | 4 | **Empowering** : increased voice and well-being for the poor | | Long-term and flexible |

*Source:* DfID 1999, 2000, and Carney 1998,

These guidelines make an important contribution to the overall value of the approach. The approach locates people at the centre of the development process, starting the analysis from people's livelihoods and their changes. Peoples' understandings and perceptions about their own problems, livelihood resources and opportunities are placed at the heart of the analysis. Hence, the partnership, participation and empowerment are naturally generated as people are at the centre of the development process. On other hand, this concern is equally important both at macro and micro-levels. The isolation of rural areas from the mainstream process has frequently led to an underestimation of the impact of policies and events upon rural people. Therefore, the livelihood approach attempts to bridge the gap between macro-micro realities, emphasizing the importance of macro level policies and institutions to develop livelihood options at micro level.

As Hussein (2002) suggests, holistic nature and peoples' strength are more fundamental principles in a livelihood analysis. In a holistic perspective, the approach considers the most pressing problems faced by people and the promising opportunities open to them, regardless of where these problems occur.[9] Therefore, livelihoods are built upon the definitions and perceptions of local people about their opportunities and constrains where feasible. The SL approach therefore supports people to realize their desirable goals. Therefore attempt is made to recognize individual's capacities to remove the constraints and to realize livelihood potentials. These enabling capacities and potentiality of local resource context are more crucial to overcome livelihood vulnerabilities (see below). Therefore, the approach guides people to become more robust, stronger and better able to achieve their ultimate livelihood objectives.

However, as Toner, (2003) argues that these principles are not a set in stone or rigid guide lines. These principles have been adopted and changed according to the executed programmes, resource availability and institutional capacity of different agencies. Marzetti (2001) attempts to translate the original guiding principles into the Brazilian context, with new principles covering gender, rights and poverty as necessary (Carney, 2002). However, these set of principles is used as an essential checklist of current best practices, and they guide how livelihood activities are effectively designed, implemented and evaluated.

## Sustainable Livelihood Framework (SLF)

The SL framework as illustrates in the Fig. 5.1, is a schematic presentation of the key elements that affect people's livelihoods.

The whole idea about the sustainable livelihoods is synthesized in this framework and helps to understand and analyze the complex relationships among different factors that are crucial for a secure livelihood system. This is however a simplification of livelihood system and it does not attempt to provide an exact representation of reality. But, it endeavours a rational way of thinking about the livelihood systems of people that will stimulate debate and reflection, thereby improving performances of poverty reduction (DfID 1999).

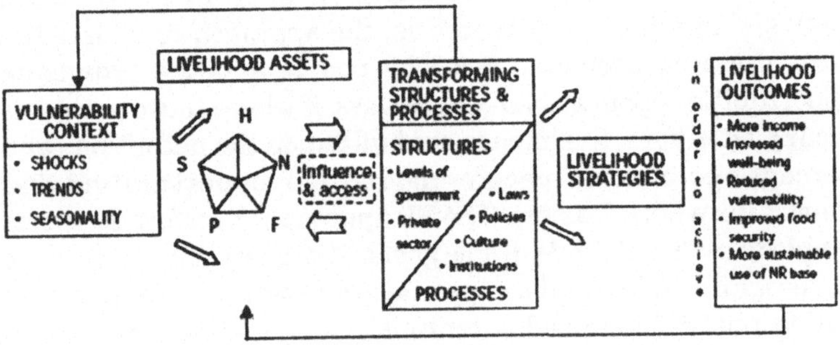


**Fig. 5.1.**

*Source:* DfID Guidance Sheet

This chapter refers the DfID's livelihood approach and framework acknowledging this is the one mostly adopted to design, implement and evaluate people's livelihoods. However, there are number of other frameworks can also be found among the different agencies that employ livelihood approach for their development interventions (e.g. CARE, IFAD, FAO, and Khanya). These frameworks provide 'essential' checklist for understanding livelihoods of people rather than a rigid format for the analysis. The DfID framework and other institutional frameworks basically describe five major components that are important for any kind of livelihood analysis.

- Vulnerability context
- Livelihood assets
- Transforming Structures and Processes
- Livelihood strategies and
- Livelihood outcomes

These elements assist to planning new livelihood development activities, assessing, analysing and evaluating the contribution of development interventions in terms of livelihood sustainability. The following section provides a detail description about these different elements.

## *Vulnerability Context*

Vulnerability is one of the foremost characteristics of the livelihood system. However, the way it presents in the framework does not suggest that the starting point for all livelihoods is the 'vulnerability context' that through a series of permutations yields 'livelihood outcomes' (DFID, 1999). In livelihood context vulnerability is recognized as a 'constant reality' that households are embedded in (Chambers, 1995b, 1995c; Chambers and Conway, 1992; Jazairy, Alamgir, and Panuccio, 1992; Wisner, 1993; Wisner and Blaikie, 2004). So often, the present livelihood analysis adopts (Chambers, 1989)'s definition of vulnerability that, refers to

> *'exposure to contingencies and stress, and difficulty in coping with them'; vulnerability has two sides: an external side of risks, shocks, and stress to which an individual is subject; and an internal side which is defenselessness, meaning a lack of means to cope without damaging loss* (1989:1).

As Chambers, (1995b) describes, many forms of losses that, severely affect upon livelihood system are becoming or being physically weaker, economically impoverished, socially dependent, humiliated or psychologically harmed. Vulnerability is thus a concept that combines *exposure* to a threat with *susceptibility* or sensitivity to its adverse consequences.

*Trends, shocks* and *seasonality* are the main causes of the livelihood vulnerability and their relative importance for different rural locations is likely to vary. Trends for example, constitute either in economic, social or political forms and may have fortuitous or adverse effects that livelihoods can be made more or less vulnerable (e.g. changes in population, environment conditions, patterns of governance, economic conditions and technology). Shocks are having widespread impacts on individuals as well as the society in scope, and of course it challenges livelihood sustainability. Shocks are unexpected events (e.g. natural disasters, war and civil unrest, episodes of disease or illness) and they destroy livelihood assets (see below), and force people to abandon their basics, bringing adverse impacts on substances livelihoods. Seasonal shifts in prices, employment opportunities and food availability are another greatest and most enduring source of hardship for poor people in developing countries. Such seasonal fluctuations reduce and rarely increase the availability of different resources at different time of the year. As Yaro, (2004) notices however, *vulnerability* and *sustainability* are two ends of a continuum. The properties of a *vulnerable livelihood system* are contrary to those of a *sustainable livelihood system* and they are defined as '*processes*' rather thai *events* (UNDP). Hence, the understanding of multidimensional aspects of vulnerability is crucial as they determine people's accessibility to various assets and livelihood options.

## 'What people have?' - Livelihood Assets

The livelihood approach is based on the premise that the 'asset status' of the community is fundamental to understanding the options open to them, the strategies they adopt to attain livelihoods, the outcomes they aspire to and the vulnerability context under which they operate (Carney, 1998a, 1998b; Chambers and Conway, 1992; Coleman, 1990; Ellis, 2000; Moser, 1998; Scoones, 1998; Sen, 1981, 1997[10]). Therefore, 'asset' is an integrated element with other livelihood components.

In livelihood context, assets are defined synonymously to resources or capital referring to something that can be used to build a livelihood (deHann, 2000). According to Bebbington, (1999), a person's assets, such as land, are not merely means with which he or she makes a living; they also give meaning to that person's world. Therefore, assets are not simply resources that people use in building livelihoods; they are assets that give them the capability to be and to act (1999 p 2022). They should not be understood only as *things* that allow survival, adaptation and poverty alleviation; they are also the basis of agents' power to act and to reproduce, challenge or change the rules that govern the control, use and transformation of resources (1999 p

2022). Coleman (1990) for example asserts that *capital and resources are productive in the sense that they both facilitate ends which would not be attainable in their absence* (1990 p 302). Therefore, assets are recognized as the central point of livelihood analysis and are depicted at the centre of the livelihood framework. A proper understanding about the asset level enables to explore people's capabilities and importantly, to distinguish what do people really have and what they don't. The ability to pursue different livelihood strategies therefore depends on the basic material and social, tangible and intangible assets that people have in their possession. In asset analysis therefore considers how access to assets has changed over time, what changes are predicted, what the causes of changes are and how access and control of assets differs between social groups (Carney, 1998b). According to Blaikie, Cannon, Davis, and Wisner, (1994), Carney *et al.*, (1999), Chambers, (1995a, 1995b) and Chambers and Conway, (1992) there are five types of livelihood asset exist;

**(*i*) "People strengths, skills and capabilities" - Human Capital:** The OECD[11] defines human capital as *the knowledge, skills, competence and other attributes embodied in individuals that are relevant to economic activity*' (OECD, 1998). According to Harris and deRenzio, (1997), skills, training and education and healthy people are the typical features of the human capital. As Coleman (1990) notices these skills and knowledge can be built by broadening one's social relationships, or investing economic aspects or re-investing human capital (e.g. reading a book). In livelihood context, skills, knowledge, ability to labour and good health that together enable people to pursue their livelihood objectives are considered as human capital (Moser, 1998). Both the quantity and quality of labour resources available to households are subsumed under human capital by some analysts (e.g. Carney 1998). However, in relation to livelihood context, human capital is considered crucial for attaining other four types of assets. Especially, among the poorer communities, human capital is often said as the chief asset possessed by their own labour (Ellis 2000:33). At household level, human capital depends on the amount and quality of labour available and it varies according to household size, skills, leadership potential, health status etc (de Hann, 2000; DFID, 1999; Rakodi, 1999).

***(ii)*** **" Natural environment around them" - Natural capital:** In a broader sense, natural capital constitutes three major components such as renewable resources, non-renewable resources and environmental services (Berkes and Folke, 1992; Throsby, 2001).[12] As DfID (1999) defines, natural asset is composite of intangible public goods such as the atmosphere and biodiversity and divisible assets like trees and lands that use directly for the production. According to ( Costanza, 1992, and Costanza *et al.*, 2000), a minimum necessary condition for sustainability is the maintenance of the total natural capital stock at or above the current level. This resource stock can be augmented when it is brought under human control that increases its productivity, as has occurred since the beginning of sedentary agriculture with the evolution of farming systems (Ellis 2000 p 32). However, the natural capital on which the rural people most depend may, because of their lack if access to private assets, be common pool resources. Especially, in relation to natural capital land and security of tenure are major issues (Rakodi, 1999) and so often land insecurity severely affect the sustainability of the rural livelihood options.

***(iii)*** **"The support they get from the communities and families they live in" - Social Capital:** In livelihood context social capital refers to mean 'social resources' including features of social organizations such as trust, norms and networks that can improve the efficiency of society by facilitating coordinate action. According to Carney (1998, p 7), DfID (1999) and Scoones (1998:8) social capital is the *'social resources'*. These include

*(i)* *Networks and interconnectedness:* this includes either vertical [patron/client] or horizontal connections [between individuals with shared interests], that increase people's trust and ability to work together and expand their access to wider institutions, such as political or civic bodies

*(ii)* *membership of formalized groups*: this often entails adherence to mutually agreed or commonly accepted rules, norms and sanctions

*(iii)* *relationships of trust, reciprocity and exchange:* this facilitates co-operation, reduce transaction cost and

may provide the basis of informal safety nets among the poor.

Bebbington, (1999) refers social capital to *'horizontal social relationships'* in livelihood analysis focusing on trust and shared values. Chambers and Conway (1991) define social capital as a part of the claims which facilitate the people's accessibility to livelihood resources. These claims are made on individuals or agencies, on relatives, neighbours, patrons, chiefs, social groups or communities or on NGOs government or the international community (Chambers and Conway 1991: 8)

According to DfID (1999), social capital has a direct impact upon the other type of livelihood assets; because social capital increases people's income and rates of savings by improving the efficiency of economic relations. On other hand social capital enhances the management of common resources (e.g. natural assets) and maintenance of shared infrastructure (e.g. physical assets). Importantly, social capital has a close relationship with human capital as social networks facilitate livelihood innovations, development and share of knowledge. To conceive livelihoods as partly depend upon households, social capital offers a more integrated framework for thinking about access to resources. As Glavovic *et al.* (2002) argues that, linkages with strong regional or national organizations could help communities to defend their resource rights, gain access to government services, lobby for changes in government policy (Moser, 1998; Narayan and Pritchett, 1997).

**(*iv*) "The infrastructure they are able to use" - Physical Capital:** Improved access to physical or produced capital is an essential element of strategies to reduce household poverty. Physical capital comprises the basic infrastructure including transport, shelter, water and sanitation facilities, energy (e.g. electricity) and communication networks (Carney, 1998b; deHann, 2000; Rakodi, 1999; Scoones, 1998). This further includes the production equipments such as factories, machines and tools and also consumer capitals like refrigerators, deep-freezers, telephone, television, internet connection etc (see DfID, 1999, Scoones 1998, and Ellis 2000 for example). In economic term physical capital constitutes producer goods created by economic production processes and consumer goods (Throsby, 2001). According to Ellis (2000) consumer goods are purchased and consumed for its direct effects on material standards of living; producer goods are purchased to generating flow of outputs for future purposes. Items that are considered to be 'consumer durables' for high income groups, such as refrigerators or sewing machines are physical capital when they are purchased as a means of generating a future flow of income (Ellis 2000:33). Insufficient or inappropriate physical structures (i.e. poor infrastructure) constraints people's productive capacity and therefore human capital at their disposal.

*(iv)* **" Money that they can use to get more of the other assets"- Financial Capital:** This is probably the most 'versatile' and also the 'least available' asset for the poor. Simply, the financial capital refers to the stock of money or financial resources that people use to achieve their livelihood objectives (Ellis, 2000; Scoones, 1998). Although, financial capital economically robust, in livelihood analysis it uses as an important livelihood asset that enables people to adopt different livelihood strategies. This constitutes mainly two sources such as stocks and regular inflows of money. Stocks include savings in the form of cash, bank deposits or any liquid assets (DFID, 1999) such as livestock (see Dercon and Krishnan, 1996 and Swift, 1989) and jewellery. Regular inflow of money includes the various types of remittances such as pensions or other transfers from the state. But this does not include the earned income.

## *Transforming Structures and Processes – TSP*

The asset portfolio and income earning activities either in household or in community is mediated by a number of factors which are determined by social, economic and political contexts. Scoones, (1998) for example identifies two types of processes

(*i*) Conditions and trends and

(*ii*) Institutions and organizations.

Scoones includes history, politics, economic trends, climate and agro-ecological conditions, demography and social differentiations as conditions and trends. In livelihood context, institutions refers to formal rules, conventions, and informal codes of behaviour that comprises constraints on human interactions; e.g. land tenure arrangements (property rights), the way markets work in practice. These norms and behaviours persist over time by serving some collectively valued purposes as well. According to (North, 1990), *they are the rules of the game of a society.* In this context, organizations are the *players;* groups of individuals bound by a common purpose to achieve common objectives. These include government agencies (e.g. Ministries and Departments), administrative bodies (e.g. local government), NGO associations (e.g. farmers' associations) and private companies. Carney, (1998b) has denominated another two categories.

(*i*) Vulnerability context and

(*ii*) Transforming structures and processes

Carney refers *vulnerability context* to the list of conditions and trends that is defined by Scoones, including shocks and local cultural practices additionally. Transforming structures constitute institutions, organizations, policies and legislations that shape livelihoods. Carney, (1998b) describes two ways in which these transforming processes impact upon livelihoods.

(*a*) It determines both who gain access to which type of asset and what effective value of that asset is and,

(*b*) In conjunction with people's asset status, they help define which livelihood strategies or activities – natural resource –based or otherwise –are open and attractive

Chambers and Conway (1998) classify *policies, institutions, laws, incentives* and *social relations* as 'transforming processes'. Reardon and Vosti, (1995) include market facilities, prices, village and regional infrastructure facilities, productions and conservations, technologies, village level assets, poverty and population pressure as conditioning factors for gaining livelihoods. Ellis (2000) identifies *social relations, institutions* and *organizations* as mediating processes. He refers social relations to social positioning of individuals and households within society, including *gender, caste, class, age, ethnicity and religion* (2000:38).[13] However, in this context as Shankland, (2000) notices the relationships between policies and livelihoods rather mechanistic. Policy operates through specific institutions to influence people's choices of livelihood strategies. This assumes a rational choice model is being operated as people choose livelihoods strategies and hints at a direct causal link between policy and action, which potentially denies full engagement with the complexity of the linkages (Toner, 2003). In a generic term transforming structures and processes are appeared to be negative in livelihood context. But, they can reinforce positive choices as well. If those structures and processes function well, they will facilitate the mobility in labour markets and reduce risk and the transaction costs associated with embarking upon new ventures (Carney, 2002; Carney *et al.*, 1999; DFID, 1999, 2000, 2001 October).

## *Livelihood Strategies*

Taking account of the livelihood assets at their disposal, the vulnerability context in which they operate, and the policies and processes around them, households tend to develop the most appropriate livelihood strategies. Livelihood creation is a dynamic process and people combines different strategies to meet their various needs at different times. It is rather clear at the household level; because, different members of the household live and work in different places, temporarily or permanently to meet their ultimate livelihood objectives. Sometimes, these strategies may lead to more or less satisfactory livelihood outcomes depending on the conditions of the surrounding factors and their changes. Although, different livelihood activities have different requirements, the general principles is that those who are amply endowed with assets are more likely to be able to make positive livelihood choices (DFID, 1999, 2000).

In the livelihood analysis two types of livelihood strategies are defined; coping strategies and adaptive strategies. Coping strategies are short-term or temporary response to external shocks and stress (deHann 2000). Depending on the severity of the response, coping strategies fade away and normal livelihood strategies return. Adaptive strategies are long-run and with the economic, social, environmental and political trends people use their own strategies to alter the situation during their lifetimes. Such adaptive strategies result in adaptive livelihoods and then the idea of adaptation vanishes and the adaptive strategies are considered to be normal livelihood strategies. deHann, (2000), Hussein and Nelson, (1998) and Toner, (2003) for example identify three main ways that such strategies are constructed: agricultural intensification, diversification and migration. Ellis (2000) divides livelihood strategies between *natural resource* and *non-natural resource* based activities.[14] Reardon, (1997) identifies a range of non-farm activities as diversification strategies of rural households in the African context.[15] Toner, (2003) for example notices that impoverished households are compelled to use divestment coping strategies that include sale of productive assets, abandoning customary economic activities, and out-migration.

## Sustainable Livelihoods in Practice

Today, livelihood concept and approach has been employed by a number of government, non-government and multi-lateral organizations as a key instrument in setting up rural development strategies (e.g. UNDP, CARE, FAO, IFAD, WB). Today this approach is utilized as a tool for understanding the complex nature of poverty and, to assess community's Strengths, Weakness, Opportunities and Threats (SWOT) prevail under different social, economic and political circumstances (Goldman *et al.*, 2000a; Goldman *et al.*, 2000b). In this sense the livelihood approach (SLA) is frequently considered as an integrated package of policy, technology and investment strategies together with appropriate decision-making tools which can be used together, to promote livelihoods by building on local strategies. illustrates some of the recent projects and initiatives completed successfully under the livelihood approach.

**Box 5.2: Uses of Sustainable Livelihood Approach: Successfully employed development-related livelihood initiatives**

1. National level planning (e.g. Uganda's Plan for the Modernization of Agriculture)
2. DfID country programme operation (e.g. Cambodia)
3. The development of large-scale poverty eradication strategies (e.g. by Khanya in South Africa)
4. Input to Participatory Poverty Assessment (e.g. Indonesia and Vietnam)
5. Structuring a large-scale living standards survey (e.g. Mongolia feeding into WB project)
6. Responding to an emergency (e.g. DfID's approach to Orissa, India)
7. FAO vulnerable group profiling
8. Project planning (e.g. Chars Livelihoods Assistance Programme (Bangladesh), Livelihood Asset Status Tracking (India))
9. Community-based planning (e.g. by Khanya in South Africa, Nepal-UK Community Forestry Project)
10. Review of existing programmes (e.g. Pakistan: preparatory study)
11. Institutional analysis (e.g. Institutional support for Sustainable Rural Livelihoods in South Africa)
12. Policy Process Analysis (e.g. Soil Fertility Management and SL: New Approaches to the policy process, Ethiopia, Kenya, Mali, Malawi, Nigeria, Zambia, Zimbabwe)

*Source:* Carney 2002 p 18

## A. STORIES FROM AFRICA

### Livelihood security programme in Tanzania

Case study 1
**Magu District, Tanzania**

This project has undertaken by CARE International in Tanzania in (1997-2000) employing the Household Livelihood

Security (HLS) framework. Many poor households in this area had lost access to cattle, the district's most important form of security against and farm land and have forced into wage labour to survive. In such situations, improvised households were compelled to use divestment coping strategies including the sale of productive assets, abandoning customary economic activities and out-migration. The project aimed to improve livelihood security in chosen villages by

- improving access to agricultural technologies,
- improving agricultural input supply mechanism
- strengthening community based groups and self-help capacity
- promoting women's self-help groups

Participants have selected from 5000 vulnerable family households in Magu district in Tanzania with a special attention to women-headed households. In terms of addressing SL principles, the interventions took households livelihoods as the starting point for action, but disaggregated in terms of gender with the participatory exercise allowing them to identify and prioritize the problems and opportunities in their own community. Partnership has formed with the district agricultural department, commercial farm input suppliers, research organizations and community-based organization.

Case study 2

**Tanzakesho District, Tanzania**

The second project in Tanzaksho in Tanzania has designed by UNDP and conducted by a core team of district council officials in 1999. The project has implemented in 29 villages. The project has conducted as a participatory planning process for achieving sustainable development after considering the drawbacks of early top-down nature of project implementations. Planning process has designed using PRA methodology in order to reflect local needs. The project has implemented as a 'micro-project', jointly funded by villagers and UNDP through the imposition of new by-laws. The project has not adopted SL framework itself, but extensively the SL principles for project design and implementation. The project has :

- raised environmental awareness and villagers have implemented new by-laws to protect environment resources in response
- increased the spirit of community self-help and improved community creativity for solving local problems
- developed a strong inter-departmental collaboration, cross-sectoral work and good partnership between districts and villages

## Project Outcomes

Both were livelihoods oriented in that they started with a holistic view of the composition of livelihoods at the micro-level. Both projects have identified as 'successful' in terms of meeting the objectives that they set out to achieve. The Magu project for example has succeeded in terms of reported increase of livelihoods and food security of target households. Some of the activities and technologies advocated by the project (i.e. organic pesticides) and the investment on existing traditional institutions have contributed to the sustainability of livelihoods in the longer term. Even though the Tanzaksho project has not explicitly adopted the SL approach, it has fulfilled many SL principles and succeeded as aforementioned.

Despite some shortcomings these two case studies have shown many features of good practice as defined by the 'SL principles'. This success appeared to derive from a commitment to general principles of

- encouraging and responding wide participation in order for activities to respond to priorities identified locally
- flexibility with regard to the content and local adaptation of activities and
- building the capacity of community-based groups of implement activities

In terms of SLA Tanzaksho project has demonstrated support for the idea that SL principles are actually a distillation of current thinking on best development practice. Especially the Tanzaksho project has demonstrated more institutionally sustainable structure as this project worked with existing structures Rather than trying to fit into either boxes 'SL' or 'non-SL' projects have better addressed whether or not these interventions impacts on poverty in the longer term and to what extent they act as 'growth-poles'. It has concerned with the development of empowering participatory local planning mechanism. The projects were flexible and successfully drawn together a cross-sectoral team who worked in partnership with each other and with external stakeholders. Even though the clear evidences on the extent to which these projects contributed to the creation of 'sustainable' livelihoods were limited, as project conclusions explained livelihoods had improved due to the effective interventions.

## EXAMPLES FROM ASIA

**Case study 1**

**Sustainable Coastal Livelihood Project – South Asia**

The project has started in 2002 and funded by the DFID Policy Research Programme. The aim of the project was to identify and promote policy processes that support the development of sustainable livelihoods for the poor in the coastal communities of South Asia. The project has covered the following issues.

- Factors supporting and constraining the achievement of, sustainable livelihoods of poor coastal people and the influence of policy processes on achieving those livelihoods
- The role of policy structures and processes in the livelihoods
- Measures for improving the policy structures and processes to support poor coastal communities in their efforts to achieve sustainable coastal livelihoods and
- Guidelines for improving policy structures and processes needed to facilitate and support the achievement of sustainable coastal livelihoods

(*Source:* Sustainable Coastal Livelihood Project South Asia 2002 Available http://www.livelihoods.org/lessons project_summaries/Coastal1_projsum.html accessed 31.05.2007)

**Case study 2**
**Integrated Food Security Programme (IFSP)**

The project has implemented in Trincomalee district of Sri Lanka during the period of 1998 - 2004. As conflict is the key constraint to development of the district, programme focused to bring various actors and agencies actively to rehabilitate and development through understanding peoples' coping mechanisms with the prevailing conditions of a protracted war. The objective of the Programme was to re-establish livelihoods through a multi-sectoral approach aiming at

- supporting communities in conflict and at food risk to diversify and intensify food and income sources
- to improve diet and health care and to facilitate peaceful coexistence

Therefore the focus of this programme was intervention was to:

- rehabilitate, re-establish and develop productive physical resources,
- develop human resources, and
- promote institutional capacities

(*Source:* Korf Benedikt (2002 November) working paper 45.

*Available:* http://www.livelihoods.org/lessons/project_summaries/comdevIFSP_projsum.html accessed 31.05.2007 and see http://ifsp-srilanka.org/html/programme_profile.html also)

**Case study 3:**
**Increasing the resilience of poor communities to cope with the impact of climate change**

The project started in October 2004 and activities have been planned over a three year period, to conclude in September 2007. The main focus of the project is to increase the resilience of men, women and children from vulnerable communities in Bangladesh, Nepal, Pakistan and Sri Lanka to cope with, and adapt to, the impacts of climate-induced hazards. The project aims to

- Strengthen the capacity of communities and government and non-government supporting institutions to prepare and respond effectively to future climate-induced emergencies

- Develop and promote practical (technology-based) interventions to strengthen people's livelihoods and natural resource assets and
- Promote the engagement of vulnerable communities in decision making processes on climate-related adaptation strategies in order to influence policy change and increase self-sufficiency

(*Source:* Increasing the Resilience of Poor Communities to Cope with the Impact of Climate Change: Second Annual Report to The Allachy Trust, from Practical Action July 2005 to June 2006: *Available:* http://www.livelihoods.org/lessons/docs/PAClimate 2.doc accessed 31.05.2007)

## Strengths and Weaknesses of the Livelihood Approach - Strengths

One of the major strengths is that, the SL approach can be used to inform activity at both field and policy level. The approach promotes exploratory thinking around the effects caused by Policies Institutions and Process (PIPs) by helping to understand which particular components of livelihoods are most affected. Many practitioners and policy makers argue the ability of the SL approach to provide a multidimensional understanding of poverty and to 'fit in' with other development approaches (Hobley, Hussein and Montagu, 2000). Some have seen the values of SL approach as an alternative approach - a 'reality check'. Many have used SL approach as a way **to facilitate change** within institutions and organisations as well (Thomson, 2000). There is a substantial body of work on the impact of policy on concepts related to SL, such as food security and poverty, which has a great deal of relevance to assessing policy impact and appropriate content; followings are some examples,

- Assessment of Macro-Micro linkages on poverty alleviation (UNDP, 2003),
- Sub-Saharan Africa's Poverty Reduction Strategy Papers (Thin, Underwood, and Gilling, 2001) and
- Participatory Living Standards Assessment 2000-Mongolia (Mearns, Dulamdary, and Shah, 2000).

- Sustainable Livelihoods, Environmental Security and Conflict Mitigation in South Asia.[16]

The recent lessons emerged from DfID, UNDP, CARE and a number of country based NGOs and local development authorities have shown success stories of SL approach in the field. For example the as mentioned earlier in the Tanzanian case studies (Magu and Taznakesho districts) both SL approach and SL principles have succeeded in securing livelihoods of poor communities (see Toner, 2003, Toner and Franks, 2006). The case study *Lekgophung* South Africa for example shows that after introducing small rural business as a secure livelihood strategy the average household income of the area has boosted by about R 3,150 (1 USD = 7.14389 ZAR as 2007) and overall disposable income by more than 26%.[17] Besides those, the community forestry programme in Nepal and the Watershed development project in India are some of the examples that succeed meeting livelihood objectives (see Carney, 1998b, 2002) and http://www.livelihoods.org). Since its overall gravity towards grassroots and its people, the SL approach has been succeeded as a tool for evaluating socio-economic impacts and the effectiveness of particular projects and programmes in poverty alleviation (Ashley, 2007; R Bond and Mukherjee, 2001; Turton, 2001).

Secondly, the approach draws attention implicitly to the assets and strengths, 'what people already have' rather than 'what they don't have'. In its holistic nature the approach attempts to get a realistic understanding of 'what shapes people's livelihoods and how the various influencing factors can be adjusted'. Therefore, the major emphasis is given on what resources or combinations of resources are important to build their 'livelihoods' in a sustainable way. This is important for designing support activities that build on the strengths of the people. The Coastal Livelihood Programme in South Asia for example, has focused on the coastal resources available for the costal people and priorities have given to remove institutional barriers enhancing their resource accessibility. The Livelihood and Forestry programme in Nepal for instance focuses on to enhance the assets of rural communities through more equitable, efficient and sustainable use of forest resources (http:/

/www.lfp.org.np/). On other hand livelihood approach encourages 'asset management' at household levels helping to draw out different relationships among other livelihood resources. SL approach to drought and water security in Ethiopia for example, emphasized the patterns of water use and how water is combined with other assets to generate household incomes (e.g. watering of livestock; small-scale irrigation).[18]

The third point is crucial when comparing the SL approach with many of the conventional development initiatives. The realization of poverty and well-being is exclusive to the SL approach. But, poverty is no longer a question of low income; it includes other dimensions such as lack of human capital (e.g. bad health, illiteracy), poor social capital (e.g. lack of social relations and family kinship) and importantly the state of vulnerability and feeling powerlessness (disempowerment) as well. For example, although the IRDP is said as more similar to the SL approach, livelihood approach builds upon the strengths of IRDP without falling into the 'traps' caused its downfall. Because of the priorities given to people, their existing strengths and constraints, recognition of risk and vulnerability aspects of poverty and the linkages between theoretical and practical aspects SL approach become much attractive than the IRDP. Therefore, the SL approach facilitates a holistic understanding of the underlying cases of poverty at different levels which, directly or indirectly determine or constrain people's access to assets and thus their livelihoods portfolios. It filters the poorest from the poor and examines their active participation as decision makers in shaping their livelihoods.

Finally, the approach reinforces the synergy among all aspects of livelihood system[19] with a broader institutional context and attempts to disclose the problems confronting and the opportunities available for the community. The approach implicitly recognizes the necessity to investigate the relationships between different activities that constituted household livelihoods, which in turn requires attention both to intra-household and to extra-household social relations. This enables the possibilities of developing livelihood strategies (coping and adaptive strategies) according to resource availability and the way people conceived their livelihood objectives.

## Weaknesses

Considerably there are number of weaknesses could also found in the SL approach. Most of the development agencies currently and development researchers those currently work with the SL approach are confronting one major problem. That is, the difficulties of translating 'SL language' as appropriate into their institutional context and the awareness raising and capacity building of SL terminology among the staff members and key partners. On other hand, handling SL principles within their existing agenda have also been repeatedly noticed as problematic. FAO, UNDP, CARE and Oxfam (GB) encounter this problem commonly. They have observed the problems of

- Raising awareness and building capacity in SL principles and methods among staff and key partners
- Translating their substantial experiences in the use of participatory methods in field projects into an effective tool for the application of SLA
- Developing multidisciplinary team and decentralized offices and country-level field programmes

Besides, UNDP notices the 'lack of clarity' as to how sustainable livelihoods and UNDP's Sustainable Human Development (SHD) approaches fitted together during their national programme implementations.[20] As WB has experienced a centrally-driven framework or approach for sectoral intervention specifically a decentralized organization like the World Bank would not be further effective. SIDA for example notices two major failures of the current SL approach by comparing UNDP, CARE and DFID. As SIDA said, none of them have addressed (a) the issue of how to identify the poor people and (b) the way livelihood resources and their livelihood opportunities are locally distributed.[21]

However, the major issue of implementing the approach at grassroots level is the extent to which the SL 'jargon' can be understandable to 'local communities'; to what extent communities become confidence with the approach as a flexible and diverse mechanism to address their own problems and solutions. Rewording and deconstructing the heavy theoretical context of the SL approach is much needed in order to realize the approach as more people-centred, participatory and partnership. On other hand, it is important to notice that one of the dangers of this type of approach is that by representing the reality and complexity of a livelihood system in a simple and logical way, the relative importance of some factors and relationships between the factors are lost (e.g. local values, knowledge and culture).

Besides the problems encountered at agency level, there are number of drawbacks have noticed in individual researches on sustainable livelihoods. Macqueen, (2001) for instance claims that SL approach is inoperable unless livelihood assets can be directly compared. As DfID (1999) believes 'at a generic level there is no suggestion that we can - or should - quantify all assets, let alone develop some kind of common currency that allows direct comparison between assets' (DfID, 1999). According to Macqueen, asset measurements in poverty alleviation are crucial. Noticing this issue Macqueen argues that 'while it may not be possible to develop a quantitative common currency for all assets, without some means for direct comparative measurement between assets the SL approach becomes entirely inoperable' (2001, p 1). Concurring with Macqueen, (Toner, 2003) also notices that defining and measuring assets is problematic and definition of single assets have a tendency towards reductionism and fail to capture linkages and trade-offs between assets. According to (Murray, 2000) equating 'assets' theoretically with varieties of 'capital', through the 'asset pentagon', intellectually distorts the understanding of capital and politically distorts the understanding of the causes of poverty. On other hand, the status of assets held by individuals or groups can be used to measure social exclusion, and this 'dark side' of livelihood assets especially like social capital is often overlooked. The difficulties to distinguish between personal and common assets are also one of the serious shortcomings of the identification of the livelihood assets.

Bebbington, (1999), Cahn, (2002), and Glavovic *et al.*, (2002) claim, that lack of cultural aspects and historical relation of livelihood construct as a serious problem. As Adato and Dick (2002) notice, the inadequate explanation of market relations is a serious deficiency of the current SL approach.[22] They suggest that the SL approach must be used in conjunction with concepts, tools and methods of analysis that have been used in other fields such as development sociology, anthropology and history (2002). Neefjes, (2000) encounters the difficulties of implementing SL approach at the field level stressing that 'such a framework can only be employed as a tool, and does not constitute as an approach itself'. For Carney (1999) sometimes, the SL approach is over ambitious and offers insufficient practical guidance on the way forward. As found with the livelihood assets, Toner (2003) further notices the problems in standardizing the application of sets of livelihood principles. As further notices the institutional analysis is too narrow in the current livelihood approach. It concerns with only formal institutions and there is a need to broaden institutional analysis beyond governance, to include community and familial structure, norms, and the process of 'institutional bricolage', through which new institutions evolve using elements of existing social and cultural arrangements.

## Conclusion

The objective of this chapter was to provide a thorough understanding about the Sustainable Livelihood Approach (SLA) with its theoretical background. The livelihood approach has evolved as a response to the top-down development approaches. The approach has identified as a remedy for many of the Eurocentric development theories inviting people into the mainstream development process. The approach allows people to define their own problems through their experiences while enabling them to seek solutions through their own capacities. The approach has admired due to its holistic nature with more flexible, dynamic, multi-sectoral and people-centred guiding rules. The approach emphasises the importance of the concepts like vulnerability, livelihood assets and policy environment as crucial in livelihood analysis.

Today, the approach has become more powerful tool for project planning, implementation and evaluation. Many of the development agencies have incorporated the SL approach and its guiding principles with their main development agendas. Many of them have proven the approach works well both at field and the policy level. Even though there are number of strengths noticed, the SL approach has been considerably criticised. Absence of the historical and cultural aspect, market relation and gender consideration, problems with asset measurements and institutional analysis the approach still opens to further researches and field level experiments.

## NOTE

1. This type of development debates are entirely an 'economic interpretation' and dominated by the economists even though some sociologists and geographers work in the field of the social and spatial unevenness of development.
2. This theory argues that once differential growth has occurred, internal and external economies of scale will perpetuate the patterns.
3. Jobs and employment normally cannote one particular activity performed formally for a payment. It manifested by a contract, between an employer and employee.
4. Stockholm Environmental Institution (SEI)http://www.sei.se/index.php?page=overview accessed 25.03.2007.
5. International Institute for Sustainable Development http://www.iisd.org/economics/poverty/livelihoods.asp acessed 03.06.2007.
6. Oxfam (GB) www.oxfam.org.uk
7. UNDP, www.undp.org/sl
8. *Research institutions*: IISD, IDS, CANARI, International Institute for Environment and Development (IIED), Natural Research Institute (NRI), ODI, [2] *Multilateral Agencies*: FAO, IFAD, UNDP, WB, WFP, [3] *International NGOs*: CARE, Khanya, Oxfam (GB), Save the Children (UK), Society for International Development (ṢID), Development Alternative Group, [4] *Organizations exploring the use of livelihood approaches*: European Commission (EC), The Population Council, SIDA, Traidcraft, [6] *Other country based institutions and NGOs*: BIDS and Samata (Bangladesh), BSF (Belgium), CEPAL (Caribbean), CICAFOC (Central America), CLADES, and MAELA (Latin American countries), ENDA (Senegal), GRET and IRAM (France), IDM (Tanzania), IUCN, NLC (South Africa), ODG (UK) (Check for abbreviation),

9. e.g. in which sector, geographical space or level, from the local through to the international.
10. Sen's capability approach gives a similar idea. Capability means that being able to perform certain basic functioning to what a person is capable of doing and being. It includes for example to be adequately nourished, to be comfortably clothed, to avoid escapable morbidity, and preventable mortality, to lead a life without shame, to be able to visit and entertain ones' friends, to keep track of what is going on and what others are talking about (Sen 1987).
11. Investment in Human Capital is at the heart of the strategies in OECD countries to promote economic prosperity, fuller employment and social cohesion.
12. Renewable resources (e.g. fish, wood, drinking water that are produced and maintained by the processes and functions of ecosystems), Non-renewable resources (i.e. oil and minerals that are extracted from ecosystem), Environmental services (i.e. maintenance of the quality of the atmosphere, climate, operation of the hydrological cycle including flood controls and drinking water supply, waste assimilation, recycling nutrients, generation of soils, pollination of crops, provision of food from the sea and the maintenance of the genetic library). In addition, (Robert Costanza, 1992) identifies two types of natural capital including renewable or 'active' natural capital and non-renewable or 'inactive' natural capital.
13. These factors may be of fundamental importance in some rural settings. But class is an ambiguous social construct difficult to generalize across different cultural contexts. While it has a precise meanings for Marxist social analysis, its manifestation in rural societies may be by reference to partial indicators like overt difference in land ownership, personal wealth, and educational attainment across different rural social strata (Ellis 2000:38).
14. Natural resource based activities: collection from woodlands and forest, food cultivation, non-food cultivation, livestock keeping and pastoralism and non-farm activities such as brick making, weaving, thatching. Non-natural resource based activities; rural trade, other rural services manufacture, remittances and other transfers such as pensions etc.
15. (a). employment in the rural non –farm labour market (RNFLM), (b). Self-employment in the local non-farm sector, (c) employment in the migration labour market (d) employment in the farm labour market.(See Reardon 1997:735)

16. http://www.livelihoods.org/lessons/project_summaries/nres_4IISD_projsum.html
17. Source: http://www.livelihoods.org/post/Docs/Lekgophung.doc
18. See http://www.livelihoods.org/info/docs/Calow2.rtf
19. Livelihood assets, Livelihood strategies, and influences of policy, institutions and processes and peoples responses toward their changes
20. For example the Malawi SL programme was found complicated and a review of the SL Global Programme by Wanmali and Singh propose the need of (*a*) national ownership of SL programmes rather than reliance of external experts (*b*) long term partnership between national, local and external actors (*c*) a wide range of local organizations working alongside key government ministries and others with clearly defined roles and responsibilities (*d*) sufficient time for programme design and implementation and (*e*) inclusiveness and openness when promoting SLA so as build widespread, sustained support within an organization.
21. For this, SIDA suggests source to overcome such problems proposing a number of possible ways of identifying the poor people. For example, geographical area where poverty is known to be widespread, poverty line, and wealth ranking.
22. A study undertaken by Dorward and Pool, Morrison (2003) elaborate the lack of *Market* aspect is and their relationship with poverty introducing Market, Institutions and Technology as Missing links of the current SL approach.

## REFERENCES


Arce, A. (2003). Value contestation in development interventions: community development and sustainable livelihood approaches, *Community Development Journal,* vol 38(3), 199-212.

Ashley, C. (2007). Monitoring and Evaluating Impact on Livelihoods: Il Ngwesi tourism lodge, Kenya [Electronic Version]. Retrieved 1.06.2007 from http://www.livelihoods.org/lessons/Thematic/Monitoring.html.

Ashley, C., and Carney, D. (1999). *Sustainable Livelihoods: Lessons from early experience*. London: Department for International Development (DfID).

Bebbington, A. (1999 ). Capitals and Capabilities: A framework for analysing peasant viability, rural livelihoods and poverty', . *World Development,, Vol 27*(No. 12), pp 2021 - 2044

Berkes, F., and Folke, C. (1992). A system perspective on the interrelations between Natural, Human-made and Cultural capital. *Ecological Economics,* Vol 5(1), pp 1-8.

Blaikie, P. T., Cannon, I., Davis, and Wisner, B. (1994). *At risk. Natural hazards, people's vulnerability and disasters*. London: Routledge.

Bond, R., and Hulme, D. (1999). Process Approaches to Development: Theory and Sri Lankan Practice. *World Development,* Vol 27(no 8), pp 1339-1358.

Bond, R., and Mukherjee, N. (2001). Livelihood Asset Status Tracking (LAST):A Case From Rajasthan: Impact Assessment for Sustainable Development working paper 3 [Electronic Version]. Retrieved 30.05.2007 from http://www.livelihoods.org/lessons/Thematic/Monitoring.html.

Cahn, M. (2002). *Sustainable Livelihood Approach* Paper presented at the 3rd Biennial Conference of the International Development.

Carney, D. (1998a). Implementing Sustainable Livelihood Approach. In: D. Carney (Ed.), *Sustainable Rural Livelihoods: What contribution can we make?* London: Department for International Development DFID.

Carney, D. (1998b). Implementing the Sustainable Rural Livelihoods Approach. In: D. Carney (Ed.), *Sustainable Rural Livelihoods; what contribution can we make?* London: Department for International Development (DfID).

Carney, D. (2002). *Sustainable Livelihood Approaches Progress and Possibilities for Change*. London: Department for International Development.

Carney, D., Drinkwater, M., Rusinow, T., Neefjes, K., Wanmali, S., and Singh, N. (1999). *Livelihoods Approaches Compared*. London: DfID.

Chambers, R. (1983). *Rural Development: Putting the Last First*. London: Longman.

Chambers, R. (1989). Editorial Introduction: Vulnerability, Coping and Policy. *IDS Bulletin,* vol 20(no 2), pp 1-7.

Chambers, R. (1995a). Paradigm shifts and the practice of participatory research and development. In N. Nelson and S. Wright (Eds.), *Power and participatory development* (pp. pp 30-42). London: ITG Publishing.

Chambers, R. (1995b). Poverty and livelihoods: whose reality counts? *Environment and Urbanization,* 7(1).

Chambers, R. (1995c, January ). *Poverty and Livelihoods?whose reality count?* Paper presented at the IDS Discussion paper 347, IDS Brighton, UK.

Chambers, R., and Conway, G. (1992). Sustainable Rural Livelihoods: practical concept for the 21st century. *University of Sussex, Institute of Development Studies, , Discussion paper 296 (Brighton IDS).*

Coleman, J. (1990). *Foundations of social theory*. Cambridge, Massachusetts: Harvard University Press.

Costanza, R. (1992). Natural Capital and Sustainable Development. *Conservation biology,* vol 6(1), 37 - 46.

Costanza, R., Daly, H., Folke, C., Hawken, P., Holling, C. S., McMichael, A. J., *et al.* (2000). Managing our environmental portfolio. *BioScience* 50, pp 149-155.

deHann, A. (2000). Migrants, Livelihoods and Rights: The relevance of migration in development policies. *SDD Social Development Working Paper* No 4,DFID, *London*

deHann, L. (2000). Globalization, Localization and Sustainable Livelihoods *Sociologia Ruralis,* vol 40( no 3), pp 341-365.

deHann, L., and Zoomers, A. (2005). Exploring the frontier of Livelihoods Research. *Development and Change,* vol 36(no 1), pp 27-47.

Dercon, S., and Krishnan, P. (1996). Income portfolios in rural Ethiopia and Tanzania: Choices and Constraints. *Journal of Development Studies,* vol 32(no.6), pp 850-875

DFID. (1999). Sustainable Livelihood Guidance Sheet (Publication. Retrieved 25.01.2006, from Department for International Development: http://www.dfid.gov.uk/

DFID. (2000). Sustainable Livelihood Guidance Sheet (Publication., from Department for International Development: http://www.dfid.gov.uk/

DFID. (2001 October ). Sustainable Livelihoods Guidance Sheet (Publication. Retrieved 04.05.2007, from Department for International Development: http://www.dfid.gov.uk/

Elliott, J. A. (2006). *An Introduction to Sustainable Development.* New York: Routledge.

Ellis, F. (2000). *Rural Livelihoods and Diversity in Developing Countries* London: Oxford University Press.

Ellis, F., and Biggs, S. (2001). Evolving Themes in Rural Development 1950s-200s. *Development Policy Review,* vol 19(no4), pp 437 - 448.

Escobar, A. (1992). Imagining a Post-Development Era? Critical Thought, Development and Social Movements. *Social Text*(31/32), pp 20 - 56.

Esteva, G., and Prakash, M. S. (1998). Beyond Development, What? *Development in Practice* Vol 8(no 3), pp 280-296.

Friedmann, J. (1966). *Regional Development Policy: A case study of Venezuela*. Cambridge MA: MIT Press.

Friedmann, J., and Weaver, C. (1979). *Territory and Function: the evolution of regional planning* London: Edward Arnold

Glavovic, B., Scheyvens, R., and Overton, J. (2002). *Waves of Adversity, Layers of Resilience: Exploring the Sustainable Livelihoods Approach* Paper presented at the DEVNET Conference.

Goldman, I., Carnegie, J., Marumo, M., Munyoro, D., Kela, E., Ntonga, S., *et al.* (2000a). Institutional Support for sustainable rural livelihoods in Southern Africa; Framework and methodology *Natural Resource Perspective*(49).

Goldman, I., Carnegie, J., Marumo, M., Munyoro, D., Kela, N., Ntonga, S., *et al.* (2000b). Institutional support for sustainable rural livelihoods in Southern Africa; Results fromZimbabwe, Zambia, and South Africa. *Natural Resource Perspective*(50).

Guyer, J., and Peters, P. (1987). Introduction. Conceptualizing the Household: Issues of Theory

and Policy in Africa. *Development and Change (Special Issue),* 18(2), 197–214.

Harris, J., and deRenzio, P. (1997). Missing Link or analytically missing?: the concept of Social Capital: An introductory bibliographic essay. *Journal of International Development,* Vol 9(7), pp 919-937.

Hirschman, A. O. (1958). *The strategy of Economic Development* New Haven CT: Yale University Press.

Hobley, M. Organisational change and sustainable livelihoods: what is the relevance? [Electronic Version]. Retrieved 05.06.2007 from http://www.livelihoods.org/pip/pip/org-hob1.doc.

Hussein, K., and Montagu, S. (2000). Lessons for the policy, institutions and processes dimensions of the Sutainable Livelihood Approach [Electronic Version]. Retrieved 05.06.2007 from http://www.livelihoods.org/info/docs/HARP-PIP.doc.

Hussein, K., and Nelson, J. (1998). Sustainable Livelihoods and Livelihood Diversification. *IDS working paper, Brighton IDS*.

Jazairy, I., Alamgir, M., and Panuccio, T. (1992). *The State of World Rural Poverty: An Inquiry into its Censes and Consequances*. New York: New York University Press for IFAD.

Jhonston, B. F., and Mellor, J. (1961). The role of Agriculture in Economic Developmenet *American Economic Review* vol 51(4), pp 566-593.

Lewis, W. A. (1954). Economic Development with Unlimited Supplies of Labour *Manchester School* vol 22(2), pp 193-191.

Macqueen, D. (2001). Measurement malaise - is the SL approach inoperable? NRIL, Pembroke, Chatham Maritime, Chatham, Kent ME4 4NN (available at http://www.york.ac.uk/inst/sei/prp/extdowndocs.html) (03.05.2007).

Mearns, R., Dulamdary, E., and Shah, M. K. (2000). *Participatory Living Standards Assessment 2000: Summary report prepared for Donor Consultative Group Meeting*: National Statistical Office of Mongolia World Bank

Moser, C. O. N. (1998). The asset vulnerability Framework: Reassessing Urban Poverty Reduction Strategies. *World Development,,* Vol 26(1), pp 1 -19.

Murray, C. (2000). Changing Livelihoods: The Free State, 1990s'. *African Studies,* Vol 59(1), pp 115-142.

Myrdal, G. (1957). *Economic thoery and Underdeveloped areas*. London: Duckworth.

Myrdal, G. (1968). *Asian Drama: an Inquiry into the poverty of Nations*. Harmondsworth Penguin Books.

Narayan, D., and Pritchett, L. (1997). Cents and Sociability: Houesold Income and Soical Capital in Rural Tanzania *Economic Development and Cultural Change,* Vol 47( 4), pp 871 - 897.

Neefjes, K. (2000). *Environment and Livelihoods: Strategies for Sustainability*. Oxford: Oxfam.

North, D. C. (1990). *Institutions, Institutional Change and Economic performances*. New York,: Cambridge University Press.

OECD. (1998). *Human capital; an international comparison*. Paris: OECD.

Potter, R., Binns Tony, Elliott A. Jennifer, and David, S. (1999). *Geographies of Development*. England: Addison Wesley Longman Limited.

Potter, R., Binns Tony, Elliott A. Jennifer, and David, S. (2004). *Geographies of Development*. England: Addison Wesley Longman Limited.

Rakodi, C. (1999). A Capital Asset framework for analysing household livelihood strategies: Implications for policy. *Development Policy Review,* Vol 17, pp 315-342

Reardon, T. (1997). Using evidence of Household Income Diversification to Inform study of the Rural Nonfarm Labor Market in Africa. *World Development,*, Vol 25(no 5 ), pp 735-747.

Reardon, T., and Vosti, S. (1995). Links between Poverty and Environment in Developing countries : Asset categories and investment poverty, *World Development,* Vol 23( 9 ), pp.1495- 1506.

Schech, S., and Haggis, J. (2000). *Culture and Development: A critical Introduction* Oxford: Blackwell Publishers Inc.

Schmink, M. (1984). Household Economic Strategies: Review and Research Agenda. *Latin American Research Review,* 19(3), 87-101.

Scoones, I. (1998). Sustainable Rural Livelihoods: A Framework for analysis. *IDS, Working Paper Brighton, IDS.*

Sen, A. (1981). *Poverty and Famines*. Oxford: Clarendon Press

Sen, A. (1997). Editorial: Human capital and Human capability. *World Development,* vol 25(no.12), pp1959-1997.

Shankland, A. (2000). *Analysing policy for Sustainable Livelihoods; Research Report* 49: Institute of Development Studies: Brighton.

Singh, N., and Wanmali, S. (1998). Concept Paper: The Sustainable Livelihood Approach (Publication. Retrieved 20-05-2005, from United Nations Development Programme: www.undp.org

Stohr, W. (1981). Development from Below: The Bottom-Up and Periphery-Inward Development Paradigm In S. Walter and F. Taylor (Eds.), *Development from Above or Below?: The dialectics of Regional Planning in Developing Countries* (pp. pp 39-71). New York: John Wiley and Sons.

Stohr, W., and Taylor, F. (1981). *Development from Above or Below? The dialectics of regional planning in Developing Countries* New York John Wiley and Sons.

Swift, J. (1989). Why are rural people vulnerable to famine? *IDS Bulletin* 20(no 2), pp 8-15.

Thin, N., Underwood, M., and Gilling, J. (2001). *Sub-Saharan Africa's Poverty reduction strategy papers from social policy and sustainable livleihood perspectives*. England Department for International Development (DfID).

Thomson, A. M. (2000). *Sustainable Livelihoods Approaches at the Policy Level*. Paper presented at the FAO e-conference and Forum on Operationalising Participatory Ways of Applying a Sustainable Livelihoods Approach. from http://www.livelihoods.org/pip/pip_macro.htm.

Throsby, D. (2001). *Economics and Culture*. Cambridge: Cambridge University Press

Toner, A. (2003). Exploring Sustainable Livelihoods Approaches in Relation to two interventions in Tanzania,. *Journal of International Development, vol 15*, pp 771-781.

Toner, A., and Franks, T. (2006). Putting livelihoods thinking into practice: Implications for development management. *Public Admin Dev, 26*, pp 81-92.

Tucker, V. (1999). The myth of development: A critique of a Eurocentric discourse. In R. M. a. D. O. Hearn (Ed.), *Critical Development Theory* (pp. pp 1 - 25). London,: Zed Books.

Turton, C. (2001). Livelihood Monitoring and Evaluation - Improving the impact and relevance of development interventions [Electronic Version]. Retrieved 30.05.2007 from http://www.livelihoods.org/lessons/Thematic/Monitoring.html.

UNDP. (2003). *Assessment of Micro-Macro Linkages in Poverty Alleviation: South Asia*. New York: United Nations Development Programme.

UnitedNations. (1995). Report of the forth World Conference on Women Retrieved 04.05, 2007, from http://www.un.org/esa/gopher-data/conf/fwcw/off/a—20.en

UnitedNations. (2000). Copenhagen Declaration on Social Development. Retrieved 04.05, 2007, from http://www.un.org/esa/devagenda/sustainable.html

Vance, J. E. (1970). *The Merchant's World: the geography of wholesaling*. Englewood Cliffs NJ: Prentice Hall.

Wallmann, S. (1984). *Eight London Households*. London: Routledge.

Wanmali, S. (1999). Sustainable Livelihoods: Lessons Learned form Global Programme Experience. Retrieved 15.6, 2006, from http://www.undp.org/sl/Documents/documents.htm

WCED. (1987). *Our Common Future*. Oxford: Oxford University Press.

Wisner, B. (1993). Disaster Vulnerability: Geographical Scale and Existential Reality. In: H. Bohle (Ed.), *Worlds of pain and hunger: Geographical perspectives on Disaster Vulnerability and Food Security*. Saarbrucken: Verlag Breitenbach.

Wisner, B., and Blaikie, P. (2004). *At Risk: natural hazards, people's vulnerability, and disasters*: Routledge.

Yaro, J. A. (2004). Theorizing food insecurity: building a livelihood vulnerability framework for researching food insecurity. *Norwegian Journal of Geography*, vol 58, pp. 23-37.

# 6

# Globalization, Governance and Epidemic Disease*

*Deane Neubauer*

## Introduction

This paper places the function of public health—that social activity most directly charged with the control of epidemic disease—in a historical perspective by outlining five stages in the evolution of the western (liberal) state. It locates public health and public policy within this development. The fifth, neoliberal stage is heavily implicated in contemporary globalization. I explore some levels of governance through which globalization operates.

The ideology of the contemporary neoliberal state has become the defacto ideology of globalization as well. Neoliberal globalization is heavily implicated in the production and reproduction of both infectious and non-communicable disease. I discuss the plight of public health in a governing context in which its writ expands even as its resources are constrained. I conclude with a set of propositions that may usefully lead to further discussions about how governance may need to be fashioned for public health to meet the challenge of its increasing and expanded change to manage epidemic disease.

In raising these questions, the reader will recognize that I touch on many themes already discussed in this conference especially in the paper commentaries.

---

*Presented to the International Workshop on 'Epidemic Diseases, Environmental Changes and Global Governance in the Historical Perspective' (Keynote Address) Datte: 9th-11th December 2005, Osaka International House, Osaka (Japan.)

## Begin with Governance

The governance notion is not uncomplicated. One can say of it that to govern is to establish a relationship between leaders and led in the context of notions of authority that allow groups within a society to interact in a predictable and regulated manner. (Dahl, 1965) Within this construct we can hastily recognize that governance can be good or bad as experienced by the governed, consensual or imposed, its authorities either legitimate or illegitimate, and accordingly likely to be enduring or episodic. Our current notions of governance are framed significantly by those of democratic governance. In the sweep of human history, the space occupied by democratic governance—ideas that those who hold and wield authority do so with the consent and sufferance of the governed—has been but a brief moment. (Cnudde and Neubauer, 1969). To seek global governance of any aspect of human behavior is one task; to seek good governance is to project a particular set of values on governance.

Our contemporary, accepted notions of democratic governance, and the systems of suffrage that emerged to sustain them, grew out of historic liberalism in which the notion of the modern state itself was viewed as deriving its legitimacy from the consent of the governed. This has increasingly become the standard for "acceptable" state forms and behavior. The "international community" increasingly promotes "free and fair" elections as the defining criterion of a democratic state. The idea permeates most of our ideas about governance.[1]

The collective commitment to create and sustain this global community is a result of the profound understanding that in urgent ways the world has become a far more independent place in which the actions of one or several actors have implications for the system as a whole. It is this complex system of interdependence that we have, since the 1960's, come to call "globalization" and mean by this term a set of identifiable dynamics, processes, and outcomes—all characterized by increased interdependence.[2]

## Levels of Governance

Governance may be said to exist meaningfully on at least six levels. All of these can be directly linked to the dynamics of globalization

Governance at the global level. The global can be understood as a collection of entities and relationships that exists transnationally throughout the world. These include existing governance organizations such as the United Nations and all its components, other transnational non-governmental organizations, such as the World Health Organization (WHO) and the World Trade Organization (WHO). These organizations are constituted out of formal participation agreements by member governments. They are supplemented by an increasing array of non-governmental organizations (NGOs) that seek to act globally, often perform government-like functions, but do so with the sufferance of the government receiving their services. Global governing organizations also employ and deploy the services of non-governmental organizations to assist their own activities. Central to these activities is a relation to the public sector and its presumed functions.

Perhaps the most important agents of contemporary globalization are transnational corporations that are "located" within an existing nation state, but operate throughout the world, accounting for over 60% of global trade, and producing most of global wealth. TNCs dominate the production, consumption and financial activities of the globe to varying degrees of regulation, depending on where they operate and are the central component of the "private sectors" within most national economies.[3]

Globalization is producing a set of readily identifiable impacts, including those that affect health. These impacts diffuse, often in quite unpredictable ways, throughout the global system producing unanticipated consequences. The "production" and management of epidemic diseases is one such consequence of globalized dynamics.

Some notions of global governance envision an entity or set of entities empowered through the abdication of some elements of state sovereignty to conduct governance functions for all members—a United Nations with effective and compelling power, for example, or world courts with sanctioning power. Other approaches to global governance, however seek a more limited approach that would control and regulate the effects, outcomes, and impacts of globalization. The issues ar egiven below :

- **Governance at the regional level.** The world is increasingly global, but also powerfully geographic, and regional in orientation and behavior. The successful organization of the EU is the most celebrated recent example of positive integrative action and a willingness to imagine new constructions of state sovereignty. Regional trade agreements such as the North American Free Trade Act and the more recently debated Inter-American Free Trade Act are others. Various environmental conventions involve regional governance agreements. Table 1.1 In part seven explicates some ideas of regulatory regimes, most of which have a regional basis.
- **Governance at the Nation State Level.** Our most familiar and important governance experiences occur at the national level, a result of the organization of the world into this state form over the past four centuries.[4](A form replicated in compressed fashion in the post-colonial period.) Within "strong" nation states (meaning those with successful programs of economic development) we have grown so accustomed to national governance that we take it for granted as a constant in our "governance lives". However, from the perspective of "weak" or "failed" states, the legitimacy of the state form at the national level is itself challenged, as it also is where "nations within nations", e.g. Kurds in Turkey, Iraq and Iran, seek independent states. Even within the well-established nation states, such as those of the OECD countries, the *governance function* is highly problematized, albeit well-institutionalized, as legislatures and courts at national and sub-national levels contest what governance processes and outcomes rightly lie at the national level and which belong somewhere else in the governance equation.
- **Governance at the Sub-national Level.** Most nation states consist of smaller administrative units, e.g. states, Lander, provinces, prefectures, districts, etc. Even in well-developed states, Britain being an example, governance functions are dynamic and subject to change, often in direct response to the impacts of globalization dynamics. The de-centering of some governance functions to local

legislatures in Scotland and Wales, as well as constant negotiations about the proper roles for London city government are all cases in point. In more tenuous states, e.g. the Philippines and Columbia, the ability of the central government to establish effective authority on various sub-national regions is questionable and in some cases non-existent

- **Governance at the Local Level.** There is one national government in the United States, fifty state governments and perhaps over 40,000 local governments. China has 666 officially designated cities, 11 of them with populations over 2 million, 23 with populations between 1 and 2 million. (China Today, 2005) Rapid urbanization throughout the world, a direct consequence of globalization, has made local urban governance a matter of central concern as often the pace of urbanization outstrips the capability of urban government to provide the bare necessities commonly associated with effective governance. Further, and especially throughout Asia, national government has been reluctant and remiss in delegating either the authority or the means to deal with the demands of rapidly growing urban populations. Within urban centers, especially those defined by extreme poverty, governance issues are central as they affect such essential issues as the privatization of water supplies. In the absence of such minimal provision, disease flourishes. Who makes decisions for whom and under what circumstances, become critical questions of governance.
- **Governance at the Individual Level.** Notions of individual or self-governance may appear odd in this context, but they are essential to notions of a limited state, to a well-regulated society and to effective public health. (Peterson and Lupton, 1997) What functions and activities the state "leaves" for citizens is the essence of what comes to be "civil society", within which appear the huge array of private associations ranging from religious establishments to trade associations and through which the state and society "assign" a large array of social functions and behaviors. In the United States "private not for profit"

organizations constitute a social sector onto themselves. Indeed, global scholars see the growth or decline of civil society, of which such organizations are a marker, as one of the key indicators of the relative "health" of the state.

From this brief discussion of levels of governance it is clear that "what we mean" by governance is rarely clear without further specification. Thus, to proceed to our topic inquiry, relating globalization to governance to epidemic disease requires us to be mindful of these distinctions between who is being governed over what things by whom at what level of social, political and economic aggregation.

## A Brief Detour to the Nature of the State

Linking epidemic disease to governance engages conversations about the distinctions between what a society regards as public (as in public health) and what private. This distinction in turn is better understood in the context of the evolution of the contemporary state forms through which globalization is developing. The modern notion of the "public" from which the concept and practices of public health spring, arises out of the historical evolution of the liberal state over the past three centuries.[5] This evolution proceeds from its minimalist forms to contemporary levels of complexity. Author will illustrate these with the American governing experience. Other liberal, free market states, pursued different but related courses.

The first stage, roughly from the formation of the national state to the Civil War, has been characterized as the *minimalist state*, sometimes as the competitive state. In this period of early capital formation, government regulation of society was limited to providing for basic rules for the conduct of the nation's business, e.g. national defense, foreign policy, control of coastal and inland waterways, and a minimal legal structure for the conduct of commerce, e.g. a national currency, rules governing customs and duties, and most importantly, rules prohibiting restraints on inter-state trade. In such a society, social units were small, wealth was relatively equally distributed, the social hierarchy relatively flat. Government intervention in social life was limited. Civil society was large in relative terms. Public claims on society were small.

The Civil War ushered in a new state form, that of the *interventionist or regulatory state* in which the writ of the state expanded rapidly—first in the effort to preserve the union and then increasingly to control the rapid (some would suggest explosive) expansion of industrial and financial capital.[6] This period of maturing industrialization rapidly changed society through the invention and exploitation of technology, the use of new energy sources to fuel production, and the wholesale creation of a goods-oriented economy. Consistently from the 1880s onward political efforts were organized to impose regulation on a market driven society in which the excesses of the business cycle and the rapid aggregation of capital "unbalanced" society against the interests of the poor, free hold farmers and the emergent middle class.

Notions of the "public interest" were aggressively arrayed against those of private holders of wealth, including the creation of public health structures.[7] Importantly, corporate interests were targeted as excessive, as were elements of individual behavior such as drug use. Government regulation of food and drugs was sought that separated producers from consumers by licensed medical professionals (Musto, 1987). This is also the period of standards, the creation of professional associations, the creation of the modern university (Johns Hopkins and the University of Chicago) based on the German scientific model. Modern science was rapidly organized in professional education, research and practice settings and given broad reach in the search for patterned regularity (in both nature and society) and as source of "scientific" principles of social regulation, from engineering to administration. The interventionist state persisted through the crises of the stock market collapse in the late 1920's and the onset of worldwide depression.

The *welfare state* that grew out of the interventionist state built heavily on its base of hard won legitimacy for the use of the state to control capital. The welfare state vastly expanded such control through wide reaching regulation of the economy and added the revolutionary innovation of seeking to redistribute wealth outside the capitalist class through purposive government programs: for the unemployed, for bankrupt farmers, to industries themselves undone by the fundamental collapse of demand that brought on the depression. (Lowi, 1979) Much government activity, such as its public works programs, sought to kick start demand by putting workers back to work while providing overall benefit to the public interest through the creation of infrastructure of great value to the economy.

In the United States, the welfare state of the Roosevelt Administration was followed by the *positive state.* The compelling idea underlying this fourth stage was recognition that market failures were (a) probably regular occurrences within capitalist economic systems, and (b) could be immensely destructive to the entire political and economic system. The challenge was to manage them through a set of policies designed to shore up demand, effectively regulate financial institutions, and develop both fiscal and monetary policies designed to "lop off" both the tops and the bottoms of the boom and bust phases of the business cycle.[8] These efforts, a complex combination of government support and regulatory activity (including significant contributions to the business community, that some would call "corporate welfare") are part of the "managed state", a combination of market mechanisms and government/state actions taken to promote even economic development. In Europe these managed state devices became part of post-war welfare states.

The election of Ronald Reagan in the United States and the ascension of Margaret Thatcher as Prime Minister of Britain orchestrated a politically conservative attack on the welfare and positive states. In Reagan's famous terms, it was necessary to get "government off the backs" of citizens, especially tax payers. Supported by the logic of supply side economics that promised substantial economic growth by reducing the size of government revenues, both countries moved to develop a range of programs favored by political and economic conservatives, including drastic reductions in the personal income tax, reducing social program support, and seeking selective economic and environmental deregulation. In many areas including health care, market oriented solutions to social problems were to be preferred over regulatory solutions. The net result of these actions was to privilege the private sector over the public.

This whole package of policies, programs and perspectives has come to be called neoliberalism—*liberal* because it focuses on limiting the state and resolving social issues within the private sector and through markets as much as possible, characteristics of the first and second stages of the liberal state. It is *neo*-liberalism out of recognition that the older mechanisms of the first and second stages of liberalism cannot be fully recovered, nor can the more important aspects of the managed state be fully abandoned. The primary agenda of neoliberalism has been to "liberalize" economic relations, especially to promote free trade and reduce spending on expensive policies of the welfare/positive states. In the early 1990s neoliberalism had come to define much of the US political center and right, and won its most spectacular victory in 1994 with the election of a Republic party Congressional majority pledged to balance the budget, reduce government spending and cut taxes.[9]

Neoliberalism is the de facto ideology of globalization. Throughout the past two and a half decades it has been adopted in one form or another by many of the key capitalist states of the global system, from which it is projected into important institutions such as the World Bank and IMF. Structural adjustment policies (the so-called "Washington solution") pursued by both institutions during the period 1975-2000 demanded that nations receiving their assistance, especially for debt relief, reorganize their governments to contract the public sector and reduce public spending while allowing greater political and economic space for private sector activity, including the embrace of foreign capital and its primary agents, the transnational corporations. (Steger, 2000) The result world-wide, but especially in Africa, was a drastic reduction in national government programs, especially those affecting education, health and basic local service provision, such as for water and sanitation. (Szanton and Manyika, 2002)

This fifth stage of the liberal state is still much in flux. Depending on the country, neoliberal sentiment can be found throughout various parts of the political spectrum. It often coalesces around a small set of policy proposals, rather than taking form as the acknowledged political and economic program of a substantial political party (or parties). The common ingredients are tax cuts, free trade agreements, calls for selective deregulation, off-loading government activities to the private sector, and cuts in welfare programs and spending. In some cases, neoliberal policies are being proposed and followed within a particular synthetic national ideology. The turn created in China by the powerful gloss on Maoism and state communism initiated by Deng Xiaoping has permitted a complex array of market oriented private sector activity to occur without abandoning the essential ideological tropes of state socialism. In many other instances one can point to cases throughout the world of neoliberal "reform" of social policies, especially in health care and education. Common to these policies, whether occurring in Britain, France, New Zealand, Australia, Canada, Japan or even China, is the expressed need to reduce levels of central government spending. Inevitably, these are couched in terms of promoting private sector solutions for social problems; inevitably they reallocate costs from government to the consumers of such services. (For an analysis of the recent health care coverage reforms in China see, Blumenthal and Hsiao 2005. For an earlier analysis of the retreat from universal health care in Europe see, Anderson and Poullier, 1999.)

A significant literature documents the social consequences of neoliberalism. Briefly and most pointedly: neoliberal-driven globalization has resulted in a profound growth in inequality throughout the world. Much of the world is worse off in relative and absolute terms than it was thirty years ago. (Neubauer, 2000; Mulrooney and Neubauer, 2005, UC Atlas, 2005.) The implications for health, education, shelter and other dimensions of human security are stark. (Labonte, et. al. 2005) Dramatic increases in infectious disease are directly linked to these dynamics of inequality. (Farmer, 2001) The very institutions that were so instrumental in establishing neoliberal policies as part and parcel of globalization, the World Bank and the IMF, have come to agree with the UN that inequality has grown (but disagree on the causes) and that poverty in all its forms is probably the greatest human security problem of the age. (Kim, *et al.*, 2000) In many regions of the world rural economies are collapsing as urban centers explosively expand, resulting in what U.N. Secretary General Kofi Annan has termed a profound crisis of our age—the spread across the planet of urban slums that systematically degrade human life. (Annan, 2003)[10]

One can also tell a more prosaic U.S. story about neoliberalism and public health in which administration neoliberals have cut public health budgets, turned attention away from behavioral oriented research in favor of drug related interventions for behavioral linked social conditions, run off a Surgeon General who in the midst of the HIV/AIDS epidemic in the United States wanted to speak candidly about the relationship between sexual behavior and sexually transmitted disease. Scott Burris says it directly and clearly:

"The political culture has not favored public health work for a long time now. In recent years, conservative Republicans with the acquiescence and even support of moderate Democrats, have mounted an attack not only on the web of social programs within which public health work is suspected, but on the very notion of government regulation itself . . . both in style and substance..the attack is more destructive than constructive. Public health has been suffering real budget cuts for years without the impetus of a right-wing revolution," and more are on the way..The conservative plan for environmental protection and occupational safety and health is to reform what amputation is to plastic surgery. Welfare reform is cutting away the health care infrastructure for the poor, even as the entire system of health care delivery for the insured is being remade with little or no public involvement. (Burris, 1997.)

Given the trajectory of its interior logic, were the neoliberal state to play out fully as a robust fifth stage of the liberal state, one would predict a much smaller state with a relatively larger, robust private sector (and in the US model, a much larger private-not-for-profit sector), lower government spending and reduced taxes on individuals and consumption.[11] This remains the goal of self-conscious neoliberals in the US Congress for example, who self-select the label "fiscal conservative" to differentiate themselves from other kinds of American political conservatives.

The events of September 11th, however, have changed the actual and probable directions of neoliberalism and the state it seeks to form. What one can with justification term the *security state* has rushed on to the stage and threatens with its vast budgets, new governments departments focused on domestic security and foreign interventions. The result promises to be not a smaller but a much larger state, not a state with less reach, but one far more intrusive, and regulatory. Current American politics reveals the polar tensions between the neoliberal and this neoconservative state. The current huge budget imbalances result primarily from the costs of the Iraq War, which, whatever its actual origins, is being fought and rationalized by the current administration as a global battle against terrorism. Fiscal conservative neoliberals, decrying the steady accumulations of national debt, have seized on the budget crisis as a means to renew their attack on social program spending.

From the perspective being developed above, the neoliberal fifth stage state and the security state are on a collision course—one or the other will have to give. A neoliberal state would result in a smaller state with less capacity for defense, the ability to compel a domestic population and less spending. A security state would result in a larger state, continued high spending, and further incursions on individual behavior. The current extant compromise—reduce budget imbalances by further cutting social programs—is unsustainable over time.

## The State, the Public, Public Health and Epidemic Disease

I have labored the previous subject for the simple but compelling reason that public health and the state are inextricably linked, and to raise the question of governance within the context of health is necessarily to interrogate this linkage. Each stage of the liberal state has, for example, constituted new meanings of "the public". Indeed, many theorists argue that a major purpose of the liberal state is to frame a continual conversation over what the "public" means in relation to the "private" and what kinds of social activity should be assigned to each. (Best and Connelly, 1976) From this perspective the critical outcomes in each of the five stages of the liberal state have been to determine governmental regulation actions taken in the name of an identifiable "public interest" that can trump private interests. Neoliberalism is a program to shift the balance in this contest back toward private interests.

This distinction is at the heart of both the meaning and historical experiences of public health, and indeed, Peterson and Lupton's important book in the 1990's detailing neoliberal approaches to public health was called "the new public health" signifying the broad efforts within developed countries to shift health responsibility to individuals, through health promotion activity (Peterson and Lupton, 1996).

Public health in its modern form grew out of the regulatory state, part of the accepted realization of the necessary role of the state to protect its population as a whole from disease, the causes of which imputed a complex of factors and conditions some of them clearly associated with the economic activities and property rights of individuals and corporations. This insertion into the politics sphere of a compelling public interest in health initiated a turbulent and highly contested history centered on governance itself (Tesh, 1988). This history was to prove that effective public health requires the state to act in the name of a compelling public interest in contesting private interests, and for it to be sufficiently organized and resourced to effectuate its health claims on the part of the public. (That is, the state must be able to successfully induce compliance within the subject population—something that is not always the case even in developed states, witness efforts to control trafficking in drugs.) The effectiveness equation also requires that the state have within its capabilities the knowledge of the causes of disease and the capacities for successful intervention.

The writ of public health as distinguished from medicine is to reduce or eliminate incidence of disease and other causes of un-health; that of medicine is primarily to treat and to cure disease. Public health is focused primarily on issues of population health, medicine primarily on questions of individual treatment. Over the past one hundred years these have become shared boundaries as disease prevention and health promotion activities are embraced within the medical writ, and as public health often operates by organizing the tools of medical intervention, e.g. immunization, to protect large populations.

Governance enters the equation again over questions of the social production of disease, how disease affects are distributed throughout a population, and politically preferred means of dealing with disease affects. (Evans, et. al. 1994) This equation is further complicated within contemporary neoliberalism by a propensity to locate disease at the site of the individual and to prefer interventions focused on individual medical treatment or through individual behavioral change, rather than through challenging social interests that may be implicated in the social production of disease. (In economic terms neoliberalism prefers responding to the individuated demand side, rather than the population supply side of health issues.) This policy preference has been at the heart of public health policy in the developed countries over the past several decades.

The social production of disease school holds that the "causes" of many diseases lie "upstream" in the ways that production and consumption are organized.[12] In this view, the appropriate intervention is to regulate these causal forces and prevent subsequent disease. These views apply to both communicable and non-communicable disease. The politics evoked in these claims are classic to public health debates in which the presumed public interest in the prevention of disease is meant to trump the private interests organized around production and consumption (Neubauer and Pratt, 1981).[13] A significant amount of environmental politics lies within this contested terrain, especially when the putative health injury is of long and complex etiology. In the absence of definitive scientific evidence directly linking environment factors to a specific disease, efforts to regulate in the direction of population health outcomes run into fierce opposition from the affected interests (Tesh, 2000). To recap: the social production of disease schools seeks governance solutions through the regulation of organized interests; the neoliberal school seeks governance solutions through the regulation of individual behaviour.

I have elided the question of the governance of epidemics within this politics because the shifting social use of the concept to refer to the distribution and incidence of non-communicable diseases (NCDs) as well as infectious diseases has altered both the health and the governance sides of the equation. This politics was critically signaled in the U.S. Surgeon General' report *Healthy People* that conjoined in policy terms both disease prevention and health promotion and assigned them as the long-term responsibility of the public health sector. (Neubauer and Pratt, 1981). The overwhelmingly private sector oriented medical and health care establishments moved quickly to commodify health promotion and integrate it within medicine as a vehicle for reducing steadily increasing health care costs. (Kindig, 1997).

Reducing incidence through effective health promotion depends on assigning individual responsibility for health and the prevention and control of disease through a combination of medical intervention and behavior change. The report advocated this approach to controlling non-infectious disease. In line with a convention politics that seeks to mobilize social and political attention social problems by labeling them as crises, the term epidemic was attached in the 1980s to a broad array of NCD's, including cardio-vascular disease and the growing incidence of diabetes associated with obesity, and various cancers. In terms of political discourse, if one can claim the existence of an epidemic and that it broadly threatens the population, one can also claim the additional resources required to deal with the crisis it represents (Edelman, 1977).

These dynamics are currently playing out in the United States in the association between the "epidemic" of obesity and its related diseases, most particularly diabetes. The linkage of the American fast food diet to these conditions has spawned a small cottage industry of publications, but the political insulation of the food system (including the advertising and the fast food industries) from protects them from government intervention. The medical conditions may constitute a major public health problem, but public health interventions are effectively limited to social marketing tools and encouraging people to seek medical care. The past three decades have witnessed similar conjunctions with smoking, the relationship of diet to cardiovascular disease, etc. These invented epidemics of NCDs compete for policy space and resources with infectious diseases, most particularly HIV/

AIDs, and the re-emergence of older infectious disease. A case in point is the re-emergence in developed states of tuberculosis, the apparent cause of which is increased migration of global workers from the developing to the developed countries.[14]

The story is being repeated on a global scale. A combination of diseases that are sometimes called the "diseases of civilization", refer to those that emerge out of the context of improved population health and growing societal affluence, a combination that prolongs life leaving surviving populations more susceptible to NCDs. The pattern has followed improved national development throughout the world. However, increasingly in newly developed countries some of these diseases are—like their counterparts in developed countries—making an appearance in children and young adults, leading to new calls to attend these epidemic crises. This was precisely the theme in at the Beijing Forum meeting (Nov. 16-18, 2005) in which data from China were cited to detail the explosive increases of diabetes and obesity among adolescent urban Chinese.[15]

To these medically situated conditions can be added yet other epidemics clamoring for policy space, such as those drug use and abuse, sexually transmitted diseases, mental illness, sleep disorders, etc. The notion of "social epidemics" has gained fashionability out of the recent popular work on "tipping points" by Malcolm Gladwell (2000). These ideas develop political currency rapidly and affect the way that key political actors, such as legislators, view the overall health of populations.

Clearly the governance of epidemics of infectious disease, non-communicable diseases and social diseases differ. We turn to this challenge in a moment.

First, however, it is useful to note that this discursive terrain is muddied further by the almost explosive expansion of "disease" within modern society. New diseases are constantly invented or discovered, some by the so-called medicalization of society wherein elements of the life cycle are claimed within the medical model as disease; some from the expansive abilities of medicine to surveill the body and to fashion interventions based on new knowledge of the body and how it functions; and some from our continually expanding knowledge of the global microbial population and its interaction with humans and animals. Our historical conceptions of public health and its role in the world—negotiating the fraught politics that I have described within the public/private nexus and seeking governance outcomes—have been based on simpler notions of disease.

With this radically expanded canon, sorting out an appropriate and effective role for public health is daunting. Not only has the usage of epidemics radically expanded, so has the range of diseases over which the notion might be applied also radically expanded.[16] The U.S. could either be an outlying case in this phenomenon, or a frightening glimpse of our future. For example, the rise in treated disease prevalence is directly linked to obesity as the key single driver of over time health care costs, accounting for 62% of the increase in per capital health care spending over the past several years; 27% of spending increases over the past 20 years. The U.S. currently spends 50% more per capita than European countries. Eighty percent of total health care spending is linked to chronically ill patients, and the chronically ill receive approximately 50% of all clinically recommended medical care (Thorpe, 2005).

## International and Global; Weak vs. Strong States

Our dominant framing of governance in the era of contemporary globalization arises out of our experiences with the international system in place throughout most of the 19th and 20th centuries. Within it the nation is the primary actor framed within a discourse of sovereignty that separates nations, one from another, in various state forms. The system itself has been weak in some regards and strong in others, having gravitated in the latter regard to any number of voluntary efforts at cooperation and coexistence (e.g. international agreements prior to WW II on shipping, communications, and to some extent trade) and catastrophically dysfunctional in the former, resulting in two world wars, and constant background activity of minor wars. Both the celebrated attempts and failures of the League of Nations to address this breakdown and the reorganization of Europe and the Middle-East after WW I by the great powers were efforts on the part of this "system" to construct a governance function that would be effective within the prevailing constructs of sovereignty.

The efforts to establish transnational organizations of governance in the post-WW II period have followed this internationalist tradition including the UN and all its subordinate institutions and the Bretton Woods organizations. This system of nation state interaction has been eroded in various degrees by the emergence of global institutions that exist both within and outside the nation state. The paradigmatic example is the transnational corporation, which necessarily is situated legally within a nation state but operates globally in its efforts to maximize the factors of production.

Over the past thirty years the behaviors of TNC's have transformed and dominated world commerce and trade (of the 60% of global trade accounted for by the top 500 global corporations, over one-half of this trade takes place within companies of the same corporation), aggregated wealth and production in unprecedented ways, and eroded state sovereignty in important respects. As part of this transition from the international to the global, new "institutions" have emerged including a linked 24/7 equity and bond market, a global monetary exchange system that cycles about $2 trillion a day, and radical changes in the nature of corporate and national accounting (Herdman, 2002[17]).

These institutional forms may be understood as governance activity in which corporations work within, through and beyond national governments to develop a consensual governance structure. They exist cheek to jowl with other kinds of transnational governance activity such as that which regulates shared international waterways (examples include the Black and Caspian Seas,) local and regional environmental regimes, and high seas fisheries (Stokke, 2001) to name but a few. They are all part of the emergent transnational governance world that is also populated by a growing array of non-governmental organizations (NGO's) to which nation states and the regime of international regulation assigns responsibility for governmental functions (much as neo-liberalism has assigned such functions in the United States to private-not-for-profit agencies), most notably perhaps in food and disaster relief. (For a summary review of types of regulatory constructions, see Appendix A.)

Scholars of globalization argue that this erosion or slippage of national sovereignty occurs in part because developed states are really incapable or unwilling to "rein in" the behavior of their own transnational corporations, or because governments agree with the underlying premise that some vectors of erosion, e.g. trade permeability, are desirable in terms of the long run benefit that accrues to the particular nation. (Nayyar and Court, 2002) But, throughout the world, developed and developing included, these matters remain highly contested, witness the debates currently raging in many developed countries over migration (life's blood to the firms that rely on the cheap labor gained), currency valuation (and the U.S.'s increasingly desperate need to gain a revaluation of the Chinese Yuan) and the U.S. balance of payment crises.[18]

This critique of the state argues that the *dynamics* of globalization (migration and rapid urbanization, trade, labor relocation, neoliberal privatization, TNC aggregation, etc.) are responsible for various *outcomes* including persistent income inequality, poverty, civil unrest, sharp and profound crises in currency values and commodity prices—most spectacularly oil—inflation and recession, trafficking in drugs, weapons and humans, and emergent infectious and non communicable diseases that collectively weaken the policy capabilities of states. In conditions of increased interdependence, the argument goes, individual nation states are confronted with a world shaped by globalization in ways that dominate single nation politics, rendering them increasingly dependent and vulnerable to global forces (Grieder, 1997[19]).

This slippage in relative autonomy of nation states establishes a context for trying to assess the issue of state capability. As alluded to above in our discussion of public health, policy to be effective requires the capacity to implement as well as to formulate and legislate policy. Those states able to combine all three elements (especially with respect to producing sufficient resources to assure implementation) anchor one end of the continuum of state sufficiency—they are strong states. Those lacking one or more of these components—for whatever reasons—are weak (or failed) states. A recent survey of relative state capacity locates most weak states either in Africa or the former constituent components of the Soviet Union, states of the Confederation of Independent States (Sud, 2005). Weak and failed states—essentially those in which governance has broken down—are increasingly seen as threats to global security.[20] Their governance deficits dilute those of the international community as a whole (the dominant premise in the British government's recent efforts to create an effective coalition of states to provide debt relief to Africa).

To questions of regime effectiveness must be added the reality of regime corruption endemic throughout much of the developing world, most notably in Africa, Southeast Asia, India and China. Although this is not the subject matter of this paper, it is a grave error not to acknowledge how extensive corruption is in these environments and the extent to which it undercuts the implementation efforts of the state, including significantly, those in health care. For example, a recent review of healthcare in several African settings presented to a conference on "Reinventing Good Global Governance," at the George Washington University in Washington D.C. points (among other things) to the particularly heavy burden placed on ordinary citizens of marketplace restructuring for the provision and sale of pharmaceuticals within the newly liberated private sectors of some countries. Even in areas of intervention where successful experience dictates the kind of pharmaceutical intervention required, individuals often must pay up to 30% of their income for appropriate medication obtained through legitimate sources. Neoliberal penetration of the pharmaceutical markets allow as many as seventeen major firms to compete for the market. The results are not, as neoclassical theory would dictate, lower costs to the consumers, but a proliferation of non-official outlets for such drugs, which leads to rampant counterfeiting of legitimate drugs, useless for treating the medical condition. (Molyneux, 2005) Wide-spread corruption is an endemic condition of weak states and very much compromises effective policy implementation, in public health as well as other areas.

## Governing HIV/AIDs

Global experience with the HIV/AIDs epidemic suggests that in many ways it is a paradigmatic disease of contemporary globalization, and the complexity that attends seeking to govern the epidemic. The virus emerged in human populations from conditions of social dislocation in Africa (where species barriers are constantly compromised); it took hold and spread rapidly in so-called marginalized populations, leading it to be termed in many areas, i.e. Africa and the Caribbean, a disease of poverty; it is closely linked to basic human patterns of social interaction (and therefore difficult to prevent); it is a complex virus posing unique challenges to inventing a successful intervention; its treatment packages proved to be late coming and expensive to produce and administer, thereby precluding treatment for many poor patients; its human transmission routes are closely associated with transient migration, the sex industry and intravenous drug users—difficult populations to monitor and control; and governmental efforts to control the disease have often fallen victim to the destructive politics of interest group domination and the fragmentation of governmental action.

Perhaps one of the most significant facts of the current epidemic is its very magnitude. The UN 2004 Global Report on HIV/AIDS documents the epidemic's continued growth: 5 million new cases in the preceding year, and its altered major frequency to women. In the United States, for example, African-American women now account for the majority of new cases. Some countries have reported successes in reducing incidence rates, e.g. Brazil and Thailand, but worldwide the epidemic continues to grow. China and India continue to demonstrate upward AIDS trends where prevalence rates are low (0.1% in China and 1.3%), but the base numbers very large. (And as Robert Perrins has pointed out, China does not have a good record for reporting public health data that might reflect poorly on itself—these data may be underestimates.) The UN estimates that up to 10 million people in China may be infected by HIV by 2010 "*unless strong and effective action is taken.*" The highest rates of spread (but not the largest absolute numbers) are among injecting drug users. For example, among this population, HIV prevalence is estimated between 35% and 80% in Xinjiang and 20% in Guangdong. The Report indicates that in other areas such as Anhui, Henan, and Shandong, in the early 1990s the disease arose among rural people who were selling blood plasma to supplement their meager farm incomes. Infection levels of 10–20% have been found, rising to 60% in certain communities, leading to many deaths.

Because it is a disease tied largely to sexual practices and other culturally sensitive activity, its social and cultural constructions are complex and contested, as are many of the advocated preventive interventions, such as condom use. Even after twenty-five years, researchers find grave ignorance of the disease among vulnerable populations. With an estimated 5 million cases India is now the country with the largest number of people living with HIV outside South Africa. Most HIV infections in that country are acquired sexually, with an additional small proportion through injecting drug use. Vulnerable populations often have little or false information about the disease. A 2001 study in India of nearly 85,000 discovered that only 75% of the respondents had heard of AIDS, with awareness particularly low among rural women in various states. In Bihar, Gujarat, and West Bengal, only about 33% of respondents had heard of sexually transmitted infections, and only 21% were aware of links between sexually transmitted infections in general and HIV. (United Nations 2004)

Dealing with HIV/AIDs as a global disease, a national disease, or an issue of human security produces a complex and tense political environment (as has recently avian flu), because the human stakes are so large.

One major fault line lies along the traditional divide between public health and medicine: a concern with the primary responsibility between reducing incidence and the societies moral obligation to treat the sick. We commonly assume that societies can do both. Some serious thinkers have begun to question that premise in the current HIV/AIDS epidemic.

This issue was starkly drawn for example at a conference hosted by the Globalization Research Network in Tampa, Florida in April 2003. Two keynotes were presented, one by Michael D. Knox, Distinguished Professor of the University of South Florida; the other by Paul Farmer, a physician and medical anthropologist, co-founder of the Division on Social Medicine and Inequalities at Harvard and Brigham and Women's Hospital. Knox advanced the hard public health position that advocates an all out preventive effort to "get in front of the epidemic and break its back". Without this, he says, we are doomed to a long and unpredictable course of continued diffusion and infection.

"HIV is endangering the security of all people on the planet. With our scientific understanding of this illness and 21 years of experience, there should be marked reductions in new cases. There is not. It is time to reconsider the course that we are on, for this preventable worldwide epidemic shows no signs of waning..until the tide of the epidemic wanes, 100% of new government resources allocated to HIV/AIDS should be used to prevent further transmission of the disease. Leaders at the global, national and local levels are urged to make the exceedingly difficult decision to temporarily forego additional government spending for treatment and to use all available resources and influence at their disposal for prevention. To take any other course will ensure that we fail at both treatment and prevention (Knox, 2003, p. 54).

Farmer's argument constructed the case for an integrated prevention/treatment approach.

"The rapid spread of HIV demands a comprehensive global strategy that includes prevention, testing and counseling, the treatment of opportunistic infections and the use of HAART [highly active antiretroviral therapy.] Social assistance to families and communities affected by HIV is also critical. For the hardest hit communities, AIDS is merely the latest in a long succession of health threats. The greatest of these are poverty and inequality, both of which are cofactors and the consequences of HIV transmission." (Farmer, 2003, p. 53).

Author cites these arguments to indicate how alive and well are the strong public health prevention case and the social production of disease school in the HIV/AIDS debate. Were space to allow, we could see that these positions anchor a continuum and that much of its center is occupied by a range of convictions including the neo-liberal. The issue of resources is critical to these arguments. The Knox position focuses directly on the question of political will, and he sees an insufficiency of that in the world to support both massive treatment and prevention. His argument skates over that of class and marginalized populations, but the question is central to the overall epidemic. As we have mentioned on several occasions in this conference, government is motivated to act when it has a crisis, and once that crisis has passed, its interests and organizational commitments wane. This is not just a case in point for disease policy, but a generalization that holds for the policy process in general and across virtually all types of governments. And, the perception of crisis for any government is most acute when "it" affects "us". For Imperial families from Rome to China, when disease threatened the seat of government, they moved to safer ground. When disease threatened military capability, government responded with the public health tools of the day (as demonstrated by several of the papers of this conference). When the prevalence of disease threatened the capability of working class populations in the early stages of the industrial revolution, government beat back the complaints of vested interests and created the legal and institutional structures of the modern public health regime.

Governments' patience and attention are not long. When a disease such as Avian flu appears on the horizon, after a very considerable amount of social organization couched in the most dire of terms, government can be forced to act with some flourish. As Robert Perrins' paper indicates clearly, a disease certainly attracts attention when it is indiscriminate in its targets and has a high short-course mortality. HIV/AIDS lacks these properties, and is burdened by its entire history as a disease of marginalized peoples (gays, the poor, people outside the European/Anglo axis, intravenous drug users, etc.) Ambiguous status and long etiology are factors central to Knox's argument..they muddy the policy waters and let government off the hook..there is always another, more or equally important crisis to attend to.

If we revisit the governance equation so laboriously laid out in the earlier portions of this paper, we can see that the HIV/

AIDS epidemic ironically both activates and compromises the "governance function." Perhaps like no other disease in previous history, it has been responsible for the creation of a vast network of NGO's seeking to attend to the disease and its consequences in one way or another. One benefit is attention to the task—someone is on duty, doing the work. The unintended consequence is the creation of an enormously "noisy" governance "system" with literally thousands of organizations, each with a small piece of the action. With further irony, this situation allows governments to do what they do best, spend money, and derive from that act what they most seek: symbolic assurance that they are acting responsibly.[21] The question of the relative effectiveness of this ad hoc governance response to the epidemic is largely begged, represented negatively only by the data that document the continued expansion of the disease.

## Global Governance, Epidemics and Changing Environmental Conditions—some propositions and conclusions.

We return to the question posed in the introduction. When speaking of global governance, do we seek a set of institutions to which the whole of the global community might be subjected, for example through a reconstituted United Nations with new powers and new mechanisms of inducing compliance? Or, do we seek to govern elements of globalization, including elements of climate change, emerging infectious disease, etc? These are different projects, and they would perforce proceed along different pathways. To which we have added questions about the social construction of epidemics and the different approaches they might require. Even further, we have raised questions about the increasing corpus of disease conditions and entities, suggesting that many of these are socially constructed. Governing them arguably requires different approaches as well. And finally, we have suggested that governance realities owe much to the "shape the state is in", whether it is a strong state capable of concerted action, or whether it is a weak or failed state incapable of formulating or effectuating policy.

Author offers by way of approaches to these questions a set of propositions that could lead to useful further investigation.

**Proposition One**: Developing a new, generalized program of global governance (the strong governance program) is likely to occur only in the event and aftermath of a major global crisis. The League of Nations and the United Nations emerged only after catastrophic failures of the international system, in a prevailing spirit of "No more" and "Never again!" Only in the context of such manifest failure are nation states willing to trade a measure of sovereignty for the benefits of collective action. This is a rational calculation even through it emerges out of conditions of crisis and moral exhaustion. (Nayyar and Court, 2002).[222] Author has sketched this argument in the paper presented to the Hokkaido Zoonosis Conference of October 2004. In doing so I argue that it is not enough for there to be a perception of an impending crisis, the really strong regulatory program requires that the crisis already has taken place. "Crises discourse may serve to modify the content of policy talk, but for large scale regulatory efforts to take place, the catastrophe in whose name such actions are taken must already have occurred. People are only convinced of the severity of truly profound crises until their very occurrence has validated that severity. The purpose of crisis dynamics and their role in the strong governance program is to legitimize lesser efforts. It is in this terrain that global governance is emerging." (Neubauer 2005).

**Proposition Two:** Governing globalization will continue to take place along axes of "weaker regulatory regimes." The Kyoto treaty demonstrates that one or two major players on the global scene can weaken strong governance programs by a refusal to participate, as has the United States. Further, it demonstrates that even efforts that take place on the strong side of the regulatory scheme can fall short of their goals without effective means for extracting compliance from member participants. (See the *New York Times*, December 4, 2005 for a review of the gap between participant targets and achieved levels. Revkin, 2005).

**Proposition Three:** Environmental governance will proceed by establishing weak regimes of regulation. (Young, 1997). Weak regimes tend to success because agreements are forged among willing partners who share common interests in achieving mutually beneficial outcomes. Success at lower levels of regulatory activity will establish the capacity for higher-level commitments. That is, environmental regulatory governance will be incremental, discontinuous, and limited.

**Proposition Four:** The neoliberal state undercuts public health by limiting its resources at the very time its policies are increasing the public health burden. Under the neoliberal state, normal governance expressed through public health will be limited in scope and unlikely to challenge strong economic, political or social interests. Public health interventions will be focused on individual treatment and behavior change rather than up-steam causal elements of disease. Under conditions of crisis, this policy will be reversed, but when the crisis has passed, the status quo ante will be re-established.

**Proposition Five:** The security state will provide greater resources for public health activity, especially those defined by crisis conditions as of direct relevance to the security brief.

**Proposition Six:** Epidemics of infectious disease will generate effective public governance responses only when they are constructed as crises, and especially when those diseases strike quickly and unselectively across subject populations with deadly effect. Infectious disease crises in developed countries will result in broader and more resourced responses than those in developing countries. (On this proposition note the exchange between David Arnold and Robert Perrins in their discussion of Session 2.)

**Proposition Seven:** Epidemics of infectious disease that have longer periods of active on-set or are disproportionately distributed across populations external to the developed countries will result in less productive crisis dynamics and governing responses. This situation is complicated by the inability of weak and failed states to provide surveillance, treatment or to effect desired behavior change because of the weak structure of governance relations in society.

**Proposition Eight:** Epidemics of non communicable diseases or socially constructed epidemics will be less likely to effectuate crises that trigger strong governance responses. The long etiology of such diseases, the complex combinations of responses taken in their face, the survivability of many with these diseases and their distributions across social classes all contribute to lessening the combination of time-sensitive factors required for a crisis to be effective. Overtime, however, such diseases crowd the "policy space" and compete for public sector resources.

**Proposition Nine**: Epidemics of non communicable and socially constructed diseases—especially within the neoliberal state—contribute to and expand medicine as a social activity, impelled in part by the activities of the medical industrial complex, and in part by the state's preference for individuated over collective oriented solutions.

**Proposition Ten**: If unimpeded, the dynamics of globalization will continue to produce and reproduce the conditions that are resulting in the emergence of new infectious diseases, the re-emergence of older infectious diseases, and the relative increases in non communicable diseases in developing countries. These processes will be accelerated in weak states, leading to the conclusion that state building is a necessary governance precondition to dealing with the epidemics that result from these conditions.

We can expand this list by other combinations of the governance factors discussed in this paper. Governance is never a unitary thing, although it is comforting to discuss it as if it were. Governance is always situational, and driven by the extant political context. Scholars of globalization are aware that its properties and dynamics operate in ways that increase the pace and alter the nature of change throughout the world. David Harvey, an early student of changes in global capital under contemporary globalization, spoke of the collapsing of time and space brought on by the capacities for immediate global communication and cheap, rapid transportation. The world is a different place, he argued when events in one part of the world have the capacity to immediately impact other distant settings, an observation that has subsequently become a truism after the advent of SARS and the ability of other diseases to spread rapidly from one environment to another through air transit. (Harvey, 1991)

Other globalization scholars, including the venerable Peter Drucker (who would never have described himself thus), caution that the levels of interdependence of the global system may be reconstituting the nature of the system itself, bringing about change dynamics that may be wholly novel. Using complexity and chaos theory to focus the point, the suggestion is that we need to wholly re-conceptualize globalization from the perspective of its uncertain and unpredictable change dynamics. And, it is argued further; our great task must be to fashion governing approaches that account for these changes. The outlook, judging by the effort of environmental scientists to make just these arguments over the past several decades, is not promising. Our hope, once again, must be in the limited successes we have gained from regional regulatory regimes. Our task is to transfer where possible the lessons of those successes to issues of governing epidemics.

## Appendix A : Types of Governance Regimes

| *Type of Regulation* | *Goal and Properties* | *Capabilities* | *Examples* |
|---|---|---|---|
| Strong Regulatory Program | Eliminate the causes of negative health and environmental outcomes: Direct regulation of globalization | Ability to directly sanction actors in effort to eliminate cause | A world regulatory court; To some degrees the World Trade Organization |
| Weak Regulatory Program | Seek to control behavior of actors producing harmful environmental or health outcomes: Regulation of globalizations affects | Works through existing organizations, e.g. national states; requires their participation and concurrence | Various Treaties: e.g. Montreal Treaty on CFC Uses |
| Type A Weak Regulatory Program | Seeks to reform existing institutions to provide greater effectiveness including ability to engage sanctions for specific practices. | Reform NGO to produce different standards for countries at different levels of development; articulate levels of responsibility as well as entitlement for TNC's; consider elements such as international anti-trust capabilities.[1] | Target: World Bank, IMF, UNCTC, UNCTAD, the Organization for Economic Cooperation and Development (OECD), International Finance Corporation (IFC), International Centre for Settlement of Investment Disputes ICSID) |
| Type B Weak Regulatory Program: Create specific regimes of regulation[2] | Identifies specific actions in specific locations—limited targeted action | Empirical; nested in other institutional structures; "lightly administered"; form more horizontally than vertically, "complex pattern of decentralized order. | Examples would include the Great Lakes water quality regime, The Antarctic Treaty System, and the European transboundary air pollution regime. |

## NOTES

1. Authoritarian states, of course, do their business in the world quite differently. Modern fascist States like Iran allow various pretenses to public expression, but just as enthusiastically suppress it when deemed necessary. The bigger problem, by far, is that elections, even when free and fair, may not be entailing, that is, they empower leaders who then may or may not act in accordance with the expectations of those who elected them.
2. For a listing of these attributes and arguments about how they constitute a system of structures and behaviors, see Neubauer 2000, or Kim *et al.*, 2000.
3. Outsourcing, however, is radically changing how government performs activities, as private corporations are contracted to perform functions formerly directly performed by government. The use of "private contractors" by the United States to conduct important aspects of the war in Iraq, including critical security functions, is a case in point.
4. See Mark Harrison's presentation to this conference in which he reviews of the transformations in the national state system following the Vienna Conference of 1815 that concluded the period of the Napoleonic Wars and the important innovations of the new diplomacy that resulted in various international conferences to develop accords of mutual interest to the international community, including those on sanitary regulation. (Harrison, 2005)
5. Certainly non-liberal states have had and continue to have institutions that perform what we will identify as public health functions, e.g. the Soviet Union and contemporary Cuba. Socialist states arose out of a 19th century reaction to liberal capitalist states, in part to champion state performance of social functions on behalf of "the public". Many of the papers in this conference focus on the performance of public health in non-liberal state forms. Author chooses this path because it leads directly to the neoliberal state and its influences on other contemporary states.
6. This stage of the state, associated with the rapid expansion of capital to produce industrial capacity, played out differently in many countries, especially those that did not have a liberal state pretense, such as Japan during the Meiji revolution, or the Soviet Union. In these forms the State is directing the process of industrialization in part by aggregating and deploying capital. Developing a robust private sector, including civil society comes

as a reaction to overbearing state control in this first industrial stage.

7. In the United States and throughout Europe, where these debates took the form of socialist critiques of capital, they gained traction in the arena of public health through the realization even within ruling capitalist elites that a healthy working class was a necessary ingredient for continued industrial expansion as well as to provide the manpower for the projection of military power through imperialist expansion. Leslie Doyle's early work is still one of the best expressions of this linkage. (Doyle, 1979) David McCullough's work on the Panama Canal is a classic text on the linkage of public health concerns in the state through the military. (McCullough, 1997).
8. In the United States the development of the positive state quickly coincided with the emergence of the Cold War and the onset of the Korean conflict, bringing into robust existence what departing President Eisenhower would call the "military-industrial complex" in his warning against its undue influence in national policy, but also the synergy between social investment such as that for the interstate highway system that sought a supporting political coalition by rationalizing it as an investment in military relevant infrastructure (it links military bases together.) Similar rationales would be sought for expanding the government role in subsidizing university-based research and training. Seymour Melman would come to call this in the 1970's the "permanent war economy." Latterly we can perhaps better appreciate that other states created demand through state spending in the 1930s by building up their military prowess. (Melman, 1976.)
9. Ironies abound within American politics, not the least of which was the explosion of national debt under the Reagan administrations, to be echoed by the current run up of debt under the current President Bush. The Clinton administration, always cast as "left wing liberal" by Republicans, actually subscribed to many of the canons of neoliberalism, including its commitments to free trade and to the support of free trade regimes throughout the world. Clinton pointed proudly to the supplanting of military regimes by elected governments in Latin America and the development of market oriented societies and expansive private sectors. Clinton also drew repeated attention (although few appeared to pay attention) to the fact that the size of the US national government in terms of active personnel fell more during his administration than any since the early

post-war period. Much of this reduction was accounted for by reductions in the size of the active military, a source of contention for yet another faction of American conservatives, neo-conservatives who (demonstrated by the current administration) have sought the projection of American power throughout the world.

10. In his forward to the UN Habitat Report on global slums Annan predicts: "Without concerted action on the part of the municipal authorities, national governments, civil society actors and the international community, the number of slum dwellers is likely to increase in most developing countries. And if no serious action is taken, the number of slum dwellers worldwide is projected to rise over the next 30 years to about 2 billion." In developing regions, slum dwellers account for 43 per cent of the population in contrast to about 6 per cent in more developed regions. In sub-Saharan Africa the proportion of urban residents in slums is highest at 71.9 per cent, according to the report. Oceania had the lowest at 24.1 per cent. South-central Asia accounted for 58 per cent, east Asia for 36.4 per cent, western Asia for 33.1 per cent, Latin America and the Caribbean for 31.9 per cent, north Africa for 28.2 per cent and southeast Asia for 28 per cent. UN Habitat (2003) Mike Davis has termed this "a world of slums." (Davis, 2004.)
11. Reduction of taxes on personal income, and their accumulation as estate or death taxes is a central component to neoliberalism, which favors taxes on consumption, and/or flat taxes on income. Proponents view these as applying equal tax rates; opponents see them as vehicles to accelerate the accumulation of wealth and as the basis of a new class structure based on wealth. The analyses of income inequality in the US are complex and contested. For a useful examination see Weisbrot, 2005.
12. Laurie Garrett has written provocatively about how distant market forces alter local environmental conditions that in turn lead to disease increases. One of her examples in tracking increases in hemorrhagic fevers is the pressuring of the species barrier that occurs when new lands, especially those that occur when the rain forest is converted to agricultural production, bring man and animals in direct contact with microbial populations for whom the forest has stood as an ecological barrier. See Garrett, 1994).
13. Other kinds of interests are also affected, of course. Some religious commitments contest public health intervention authority to resist immunization, sex education for children, and various behavior restrictions, etc.

14. "In the United States, TB has re-emerged as a serious public health problem. In 2001, based on provisional data reported to the U.S. Centers for Disease Control and Prevention, the number of cases has decreased for the ninth straight year to 15,991 cases of active TB (infection with full-blown disease symptoms). This all-time low is due largely to improved public health control measures. In addition to those with active TB, however, an estimated 10 to 15 million people in the United States are infected with M. tuberculosis without displaying symptoms (latent TB) and about one in ten of these individuals will develop active TB at some time in their lives. Minorities are affected disproportionately by TB: 54 percent of active TB cases in 1999 were among African-American and Hispanic people, with an additional 20 percent found in Asians." The fact sheet continues on to locate the cause of the re-emergence in migration. (NIAID, 2002.)
15. The rapid growth of an affluent middle class in both China and India creates daunting public health problems as one end of the income distribution suffers from the diseases of poverty and the other from the diseases of affluence.
16. Some of the social production of these usages comes from the medical industry itself, especially the pharmaceutical industry, which is constantly seeking the expansion of categories of usage that expands applications beyond their initial use. Numerous examples of recent example of "usage creep" exist in which pharmaceutical applications are extended to healthy populations as "prevention". The wide spread use of statins to control cholesterol is a case in point, but so was the shifted application of Viagra from its original purpose. Within the medical economy this phenomenon is known as "supply pull" and is seen as a major cause of constantly increasing medical costs. Current estimates hold that by 2014 health care spending in the US will amount to 18.4% of GDP.
17. The title of Herdman's speech "Moving Toward the Globalization of Accounting Standards," signals the emergence of one of the new languages of globalization and points in yet another governance modality. At the time of the speech, Herdman was the Chief Accountant of the U.S. Securities and Exchange Commission.
18. Peter Drucker, the dean of management/business analysts who recently passed away wrote just prior to his death of the possible catastrophic consequences for the U.S. of continuing its current pattern of running huge budget and balance of payment deficits

and seeking to finance them globally, in the current moment largely through Chinese capital.

"The next major economic crisis will most probably be a crisis of the U.S. dollar in the world economy. It will put to a severe test the oligopoly of the central banks of the developed countries that now rules over the world financial economy. . . . The Bretton Woods system was never the stable, "non-political" system Keynes wanted. It could not and did not prevent currencies from being overvalued or undervalued. Still, although it limped from one crisis to the next, the Bretton Woods system worked for most of the half-century after World War II. And there was only one reason why it worked (however poorly): the commitment to it of the United States and the strength of the U.S. dollar as the world's key currency. . . . The dollar is still the world's key currency. But the Bretton Woods system is being killed by the U.S. government deficit, which is fast becoming the sinkhole of the world financial economy. . . . The government deficit is therefore being financed almost in its entirety by foreign investments in the United States, mostly in government securities like short-term treasury notes and medium-term bonds. The Japanese are converting most, if not all, of their trade surplus with the United States into dollar-denominated U.S. government securities and have thus become the largest U.S. creditor. (Drucker 2005.)

19. This has truly become a massive literature. I cite Grieder as a sharp and concise argument.

20. Weak and failed states are the incubators of many global distresses, evidence again of the interdependence element of global human security. For example, a recent study of freshwater fisheries labels them a "hidden crisis," many of which verge on collapse through over fishing and waterway pollution. Strong states can organize the means to address this situation, but weak states cannot. As population growth fuels the constant search for dietary protein, the fisheries verge on collapse; as stocks decline, fishers move on to wild land protein stock, threatening them with collapse.

21. Murray Edelman's work on the symbolic actions of government is a rich source of examples of how government acts to "fulfil" its governance expectations. See both 1977 and 1985.

24. Regimes can be understood as "sets of rules, decision-making procedures, and programs that define social practices, assign roles to the participants in these practices, and govern their interactions." (Young, 1997).

## REFERENCES

Anderson, G. F. & Poullier, J. P. 1999, 'Health spending, access, and outcomes: Trends in industrial countries,' *Health Affairs,* vol. 18, no. 3, pp. 178–192.

Best, Michael H., and Connolly, William E., 1976. *The Politicalized Economy,* Heath, Lexington, MA.

Blumenthal, David and Hsiao, William, 2005, "Privatization and Its Discontents: The Evolving Chinese Healthcare System, *New England Journal of Medicine,* Vol. 353: 1175-1170, Sept. 15, 2005, No. 11.

Burris, Scott, 1997, "Public Health Forum: The Invisibility of Public Health: Population-Level Measures in a Politics of Market Individualism," *American Journal of Public Health,* October, Vol. 87, No. 10.

China Today, 2005, "Chinese Cities and Provinces," available at: *http://www.chinatoday.com/city/a.htm*

Dahl, Robert A., 1965, *Who Governs?* Yale University Press, New Haven.

Davis, Mike, 2004, "Planet of Slums: Urban Involution and the Informal Proletariat, *New Left Review*, 26 March/April, 6-34.

Doyal, Leslie, 1979, *The Political Economy of Health,* Pluto Press, New York.

Drucker, Peter 2005, "Trading Places," The National Interest, Spring 2005, Posted On: 3/17/2005.

Edelman, Murray, 1977, *Political language: words that succeed and policies that fail*, Academic Press, New York.

Edelman, Murray, 1985, *The Symbolic Uses of Politics,* University of Illinois Press, Urbana.

Evans, Robert G., Barer, Morris L., and Marmor, Ted, 1994. *Why are some People Healthy and Others Not?: The Determinants of Health of Populations,* Aldine De Gruyer, New York.

Farmer, Paul 2001, *Infections and Inequalities: the Modern Plagues*, University of California Press, Berkeley, CA.

Farmer, Paul, 2003. "Integrating HIV Prevention and Care: the case of Haiti," in *HIV/AIDS" Prevention in a Global World,* D. Neubauer, ed., available at: globalgrn.org.

Garrett, Laurie, 1994, *The Coming Plague: Newly Emerging Diseases in a World Out of Balance*, Farrar, Straus and Giroux, New York.

Gladwell, Malcolm 2000, *The Tipping Point: how little things can make a big difference,* Little Brown, Boston.

Grieder, William, 1997, *One World Ready or Not: The Manic Logic of Global Capitalism*, Simon and Schuster, New York.

Harrison, Mark (2005), "Disease, Diplomacy, and International Commerce: The Origins of International Sanitary Regulation in the Nineteenth Century, paper presented to the International Workshop on 'Epidemic Diseases, Environmental Changes and Global Governance in the Historical Perspective," 9-11 December, 2005, Osaka International House, Osaka, Japan.

Harvey, David, 1991, *The Condition of Postmodernity,* Blackwell, Oxford.

Heffler, Stephen; Smith Sheila; Keehan, Sean; Borger, Christine; Clemens, M. Kent; Truffer, Christopher, 2005. "Trends: U.S. Health Spending Projections for 2004-2014, *Health Affairs*, 10.1377, February 23, 2005.

Herdman, Robert K., 2002, "Moving Toward the Globalization of Accounting Standards," Schmalenbach Institute for Business Administration Conference (Schmalenbach-Tagung) Cologne, Germany, April 18, 2002

Kim, Jim Yong, Millen, Joyce V., Irwin, Alec and Gershman, John, 2000, eds.2000, *Dying for Growth: Global Inequality and the Health of the Poor,* Common Courage Press, Monroe, Maine

Kindig, D. A. 1997, *Purchasing Population Health: Paying for Results,* The University of Michigan Press, Ann Arbor, MI.

Knox, Michael D. 2003. "HIV Prevention Versus Treatment: Setting Global, National and Local Priorities," in *HIV/AIDS" Prevention in a Global World,* D. Neubauer, ed., available at: globalgrn.org.

Labonte, Ronald; Schrecker, Ted; Gupta, Amit Sen, 2005. *Health for Some: Death, Disease and Disparity in a Globalizing Era,* Center for Social Justice, Toronto.

Lowi, Theodore, J. 1979. *The End of Liberalism: the second republic of the United States,* Norton, New York

McCullough, David, 1977, *The Path Between the Seas: the Building of the Panama Canal, 1870-1914,* Simon and Schuster, New York.

Melman, Seymour, 1976, *The Permanent War Economy: American Capitalism in Decline,* Touchstone Press, New York

Molyneux, David, 2005. "Global Governance and Health," Presentation to the Re-inventing Good Global Governance Conference, George Washington University, Nov. 4, 2005. Available at http://gwcsg.gwu.edu

Musto, David F., 1987, *The American Disease: Origins of Narcotic Control,* Oxford University Press

Nayyar, D. and Court J., 2002. *Governing Globalization: Issues and Institutions,* Policy Brief No. 5, United Nations University, New York.

NIAID, 2002, "Tuberculosis," *Fact Sheet,* March 2002. Available at: http://www.niaid.nih.gov/factsheets/tb.htm

Neubauer, Deane, 2000. "Assaying Globalization," *American Studies,* Vol. 41, Numbers 2/3, Summer/Fall 2000, 13 32.

Neubauer, Deane, "Globalization and Emerging Governance Modalities," 7th International Symposium for Zoonoses Control Risk Management and Global Governance of Zoonoses, Hokkaido University Sapporo, Japan 5-6 October, 2004

Neubauer, D., and Mulrooney, L, 2005. "Globalization and Health: Round Two," Paper Prepared for the Pacific Global Health Conference Honolulu, Hawaii, June 15-17, 2005

Neubauer, D. and Pratt, Richard, 1981, "The Second Public Health Revolution: A Critical Appraisal,"), *Journal of Health Politics, Policy and Law,* Vol. 6, #2, Summer, 1981, pp. 205 228.

Peterson, Alan & Lupton, Deborah. 1996, *The New Public Health: Health and Self in the Age of Risk*, Sage Publications, Thousand Oaks, CA.

Revkin, Andrew C, "On Climate Change, a Change of Thinking," *New York Times,* December 4, 2005, B-3.

Steger, M. 2002, *Globalism: The New Market Ideology*, Rowman & Littlefield Publishers, Oxford.

Sud, Inder (2005), "Promoting Stability and Development and Fragile and Failed States," The Challenge of Globalization: Reinventing Good Global Governance, November 4, 2005, George Washington University, Washington D.C., available at: http://gstudynet.com/gwcsg

Stokke, Olav Schram, ed., 2001. *Governing High Seas Fisheries - The Interplay of Global and Regional Regimes*, Oxford University Press, Oxford

Szanton, David L., and Manyika Sarah, 2002 "PHD Programs in African Universities: Current Status and Future Prospects", a *Report to the Rockefeller Foundation*, available at: *http://globetrotter.berkeley.edu/RADW/AfricaPhDReport.pdf*

Tesh, Silvia N. 1988. *Hidden Arguments: political ideology and disease prevention policy,* Rutgers University Press, New Brunswick, NJ

Tesh, Silvia N. 2000. *Uncertain Hazards: environmental activists and scientific proof.* Cornell University Press, Ithaca, NY

Thorpe, Kenneth E. 2005. "What Accounts for the Rise in Health Care Spending And What to Do About It: Future Directions in Health Care Reform," presented to Healthcare Coverage in Hawaii: Turning Point or Tipping Point, October 17, 2005. Available at: *http://www.healthcoveragehawaii.org/*

UC Atlas UC Atlas of Inequality, available at: http://ucatlas.ucsc.edu

UN-Habitat, 2003*The Challenge of the Slums: Global Report on Human Settlements*, London

Weisbrot, Mark, 2005. *Globalization: A Primer,* Preamble Center, Washington; http://www.preamble.org/globalprimer.htm

Young, Oran. (Ed.) 1997. *Global Governance: drawing insights from the environmental experience.* The MIT Press, Cambridge, MA.

# 7

# Measuring Quality of Life in Macedonia : Using Human Development Indicators

*Dimitar Eftimoski*

## ABSTRACT

By the end of the 1980s, the central issue of development was focused on the growth of income and not on the growth of quality of life. Therefore, the development strategies were mostly oriented towards production and left no significant space for improving the welfare of individuals.

In the beginning of the 1990s the human development concept emerged, stressing that economic development ultimately should result in improvement of the quality of individual's life. The main goal of the development process, according to this concept, is to expand the capabilities of individuals by placing them in the focus of the efforts for development.

This article is focused on the quality of life of the individual. Moreover, in addition to the previous practice in Macedonia of calculating the Human development index (HDI) - as a measure of quality of life, an attempt will be made to calculate the Human poverty index (HPI-2) - as a measure of non-income poverty, Gender development index (GDI) - as a measure of inequality between men and women, as well as the Human development index at the level of aggregated urban and rural municipalities. Finally, a simple qualitative analysis of the success to convert economic growth into human development, i.e. "analysis of the quality of the economic growth ", will be applied. Along with the human development indicators, in order to estimate: Gini index,

Head count index, Poverty gap index, Foster-Greer-Thorbecke index and elasticity of poverty measures in respect to the average household income and Gini index, we will employ software package "Povcal" - aimed to group distribution data.

***Key words:*** *human development, economic development, human poverty, income poverty, gender inequality, income inequality, quality of life.*

## Introduction

Immediately after the independency of Macedonia (1991), liberalization of foreign trade and price liberalization policy affected economic instability. Furthermore, restrictive macroeconomic policy was adopted. Restrictive monetary and fiscal policy, applied together with restrictive wage policy and limited banking placements, resulted in a decrease in the population living standard. The low growth rates of real GDP per capita (Fig. 7.1) could not improve people's quality of life - human development, so both inequality and poverty trends continue to increase.

## On the Concept of Human Development

Human beings, or rather identifying alternatives for the advancement of their welfare have been a primary subject of research in the economic science since its early beginnings :

> ". . . Indeed, a belief in the central importance of enriching the lives and freedoms of ordinary human beings has been a foundational concern in the social sciences for very long time. This applies not only Adam Smith, but also - much earlier - to the pioneering exploration of political economy in the 17th century by Sir William Petty. Petty's classic monograph Political Arithmetic . . . was clearly motivated by an interest in the lives of human beings in a way that is not all that removed from the contemporary interest in human development . . . "
>
> -Amartya Sen

In time, the subject of research has transferred from human beings to production. Increased production became an obsession for the economists, while the interest in the human beings' quality of life continued to decline.

In response to such tendencies, the concept of human development emerged in 1990s. This concept very clearly defined the basic goal of development – the human being, i.e. improvement of his/her quality of life.

One must have in mind that the human development concept has incredibly strong background in the history of economic thought. Undoubtedly, its roots can be found in the well-being theory, as well as in the basic needs concept. However, it was Amartya Sen's theory of development on capabilities and functioning that provided a strong conceptual foundation for Human development concept.

Human development has been defined as a process of expanding the human choice – with respect to the capabilities for a human being to have a long and healthy life, to be better educated and to have a decent standard of living. Surely, the human beings' choice is not exhausted with the above mentioned components. Particularly important are other "supplemental" components, such as: political freedom, guaranteed human rights, human security, etc., which for their part exert constant pressure for supplementing and upgrading the concept.

In a broader sense, the concept of human development could be defined as development of the people, development for the people and development from the people (Human Development Report, 1993 : 3).

The *development of the people* is directly related to human capital investments, i.e. investments in formal education, training, health care, as well as everything which directly or indirectly contributes to enhanced productivity and creativity of an individual.

The *development for the people* on the other hand is related to the distribution of income. It shows whether and to what extent the economic growth, whose generator is the human being himself, is equally distributed among the individuals. Unequal distribution of the economic growth would ultimately mean unequal distribution of the human beings' quality of life.

The *development from the people* is the last component of the broader definition of the human development concept. It is directly related to creating opportunities for people's active participation in creating their own development. Indeed, the strategies of sustainable human development also put the emphasis on generating productive jobs. For a long period of time there was a prevailing belief in the economic literature that initiating economic growth, would inevitably lead to increased employment rate. But, practice showed something different. Namely, research conducted in both underdeveloped and developed countries proved that increased output is not always accompanied by corresponding increase in the employment rate. In relation to the economic growth, the employment rate is either unchanging or it is rising less proportionately. This practically means that modern economies are facing a new phenomenon – economic growth followed by a low employment rate – *jobless growth* (Human Development Report, 1993 : 3).

The human development should be understood as a primary (ultimate) goal of the economic development, while at the same time being the most efficient means of promoting economic development.

In the first case, as an ultimate goal of the economic development, human development means improvement and enrichment of human life. Hence, it should be completely clear that the development's main goal is not production of as many goods and services as possible, but rather strengthening human capabilities for a fulfilled, productive and dignified life (Griffin and McKinley, 1994 : 1-10).

In the second case, as a means of economic development, the human development stands for the main drive of the economic development. Through the process of human capital accumulation, it enhances people's skills, knowledge, productivity and inventiveness. Thus, the economic development "profits" from the human development.

In this context, it is an indisputable fact that the real GDP growth, or increased income and employment rate, are important development components. Nevertheless, clear distinction should be made between the development goal and the means of its promotion. Income and employment rate are not development goals, but means for raising the existing level of development. The basic goal of development is to increase human capabilities and the range of human choices. Income is merely a part of the human choice. Human choice, beside the income, implies advancement of the human health, creative life in an affluent and healthy natural environment, democratic civil society, etc.

Frequently, a prevailing conclusion is that the income itself is the solution for improved human choice, i.e. that the income is the only and most efficient solution (means) for increased number of options – choices for people. However, this opinion is only partly true. Researches show that there is no automatic relation between increased income and human development (Human Development Report, 1990:10). In other words, there may be a scenario where the real income grows, while the effects of such growth do not reach people. Experience shows that there are cases of a country with high level of human development and moderate level of real income, and the opposite, a country with low level of human development and high level of real income (see Tables 7.1 and 7.2).

**Table 7.1. Different levels of real GDP per capita (PPP US $), with approximately equal HDI**

| *Country* | *GDP per capita (PPP$)* | *HDI* |
|---|---|---|
| Croatia | 9.170 | 0.818 |
| Poland | 9.450 | 0.841 |
| Estonia | 10.170 | 0.833 |
| Lithuania | 8.470 | 0.824 |
| Uruguay | 8.400 | 0.834 |
| Chile | 9.190 | 0.831 |
| Kuwait | 18.700 | 0.820 |
| Qatar | 19.844 | 0.826 |
| United Arab Emirates | 20.530 | 0.816 |

*Source:* Human Development Report (2005), UNDP.

**Table 7.2. High real GDP per capita (PPP US $) with high HDI**

| *Country* | *GDP per capita (PPP$)* | *HDI* |
|---|---|---|
| Canada | 27.130 | 0.937 |
| Norway | 29.620 | 0.944 |
| USA | 34.420 | 0.937 |
| Australia | 25.370 | 0.939 |
| Iceland | 29.990 | 0.942 |
| Sweden | 24.180 | 0.941 |
| Belgium | 25.520 | 0.937 |

*Source:* Human Development Report (2005), UNDP.

The absence of an automatic relation between the real income and human development should not be understood as a rule which unambiguously points out that the income does not contribute to an increased level of human development. On the contrary, increased income contributes enormously to the human development, but it would be wrong to link the entire human development to the income increase. There are at least two reasons for this: First, income accumulation does not have decisive influence on the fulfillment, or accomplishment of some of the options (choices) of a human being, or the society as a whole. For example, a society does not necessarily need to be rich to be democratic, or a family need not be rich to respect the dignity and rights of each member; second, the human choices reach much further than the economic wellbeing. Namely, people want to have a healthy and long life, to be better educated, to live in a healthy and clean natural environment, to enjoy the safety of their homes, workplace and society, which is of course considerably, if not completely, beyond the "range" of income.

To summarize, until the emergence of the human development concept, the central issue of development was focused on the income increase, and not people's quality of living. Consequently, the development strategies were production oriented and did not leave enough room for people's wellbeing improvement. Therefore, the quality of people's lives should not be evaluated on the basis of the average income level, but according to people's capabilities to lead healthy lives, to be educated, have self-respect and take active part in the social mainstreams. Human welfare growth is directly related to the freedom of choice. The increase in human capabilities leads to greater freedom of people's choice, and thus they can compare a large number of options that lead to increased wellbeing.

## The Human Development Level in Macedonia (situation and trends)

Before moving to the analysis of the human development level in Macedonia, it is important once again to emphasize that there is no automatic relation between the growth of the real income per capita and the human development (quality of life). The real income per capita could grow, while the effects of such growth could fail to reach people. Certainly, there are examples where the growth of the real income per capita is accompanied by an appropriate increase in the human development level (Fukuda Parr *et al.*, 2002 and Stewart, 2002).

## Human Development index - HDI

The human development index is an indicator of the average achievements in the field of basic human capabilities (human development). One has to take into account its deficiencies as well, such as the incapacity to reflect the distributive effects of the development (the inequality) and to measure the deprivations aspects of the development. Moreover, according to Sen (2000:17-24) it is important to distinguish between the use of HDI as an index and the overall concept of human development. Namely, we should be clear about the fact that there are many relevant variables of the human development that are not included in the HDI.

Human development index[2] is based on three components (the life expectancy, the achieved education and the living standard) and is not exclusively focused on the economic wealth - as the case may be with the GNP (Jahan, 2002).

The HDI for Macedonia was calculated according to the 1999 modified methodology by Sen and Anand (1999 : 1-10).

With the HDI amounting to 0.793, Macedonia belongs to the group of countries with a middle human development level.[3] According to the level of the HDI, Macedonia is number 60 world ranking out of a total of 175 countries (Table 7.4)

**Table 7.3. Components required for the calculation of HDI for Macedonia (2004)**

| *Life expectancy* | *Adult literacy rate (%)* | *Gross enrollment rate for all three educational levels* | *Real GDP per capita (PPP $)* |
|---|---|---|---|
| 73.05 | 96.0 | 70.0 | 6.470 |

*Source:* State Statistical Office, Statistical Yearbook 2004.

**Table 7.4. Human Development Index (HDI) for Macedonia**

| *Life expectancy index* | *Completed education Index* | *Adjusted (GDP per capita Index (PPP $)* | *HDI* |
|---|---|---|---|
| 0.810 | 0.870 | 0.700 | 0.793 |

*Source:* Our own calculations based on a data from State Statistical Office

High HDI in Macedonia, as well as in the countries stemming from the former so-called socialist block and Former Yugoslavia should not be surprising. We can openly sum up that in the past, these countries built up a real strong social component on the account of low economic efficiency, which today reveals throughout human development indexes. According to the World Human Development Report - 2005, in the group of countries with the highest human development level are listed even nine former socialist countries (Slovenia, The Czech Republic, Slovakia, Hungary, Poland, Estonia, Croatia, Lithuania, Latvia), whereas the others are mainly listed on the upper middle level of human development (Bulgaria, Russia, Romania, Macedonia and etc.).

The Macedonia's HDI does not confirm further the hypothesis that there is no automatic link between the growth of the real income per capita (economic growth) and the level of HDI (human development) (Fig. 7.1). The insignificant disparities between these two components mainly result from the inequality in the distribution of the income, i.e. in what way and how successfully the generated income is converted into human development.

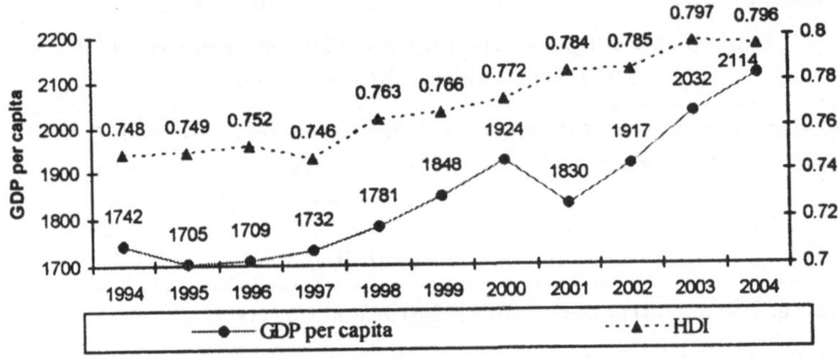


**Fig. 7.1. Dynamics of HDI and GDP per capita in Macedonia**

*Source:* State Statistical Office and our own estimates

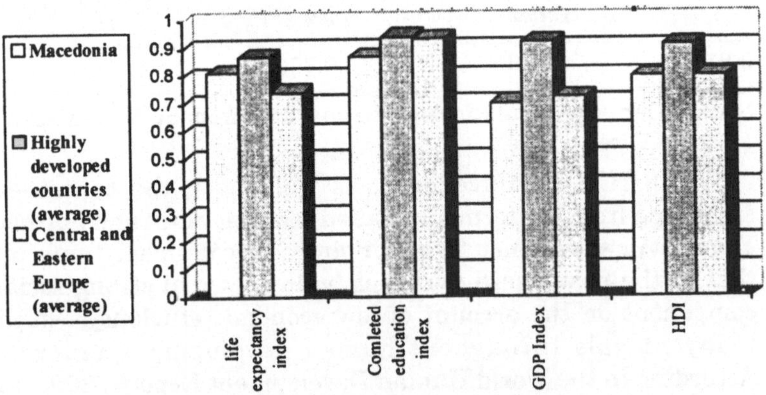


**Fig. 7.2. Comparison of HDI composite indexes with highly developed countries and the countries of Central and Eastern Europe**

*Source*: Human Development Report, 2005 and NHDR - Macedonia, 2004

The Macedonian HDI - 0.793, significantly differs from the average HDI of the highly developed countries - 0.911 (Fig. 7.2). This difference is especially significant in the "knowledge index", where the value of completed education index reaches 0.870 and 0.940, respectively. The gross rate of enrolment at all three levels of education in Macedonia is 70%, while in the highly developed countries - 87%. In addition, the income component, respectively the real GDP per capita (PPP$), indicates great differences between Macedonia (0.700) and highly developed countries (0.920). Macedonian real GDP per capita (6.470 US$) is approximately three times lower than the average real GDP per capita of the highly developed countries (24.904 US$).

Regarding the averages of HDI composite indexes of the Central and Eastern European countries, Macedonia mainly shows satisfactory results, except in the component - acquired education (Fig. 7.2).

### *HDI in urban and rural municipalities*

The disaggregated human development indexes are of particular importance for the assessment of the disparities among the municipalities (Fukuda Parr *et al.*, 2002).

The human development level in rural and urban municipalities in Macedonia is shown in Tables 7.5 and 7.6:

**Table 7.5. Components required for the calculation of the HDI per type of the municipality (2004)**

| *Type of municipality* | *Life expectancy* | *Adult literacy rate (%)* | *Gross enrolment rate for all three education levels* | *Real GDP per capita (PPP $)* |
|---|---|---|---|---|
| Urban | 73.05 | 94.81 | 78.52 | 6.418 |
| Rural | 73.05 | 90.63 | 62.36 | 5.926 |

*Source:* State Statistical Office

**Table 7.6. HDI per type of municipality**

| *Type of municipality* | *Life expectancy* | *Completed education index* | *Index of adjusted real GDP per capita (PPP $)* | *HDI* |
|---|---|---|---|---|
| Urban | 0.800 | 0.890 | 0.690 | 0.796 |
| Rural | 0.800 | 0.810 | 0.680 | 0.765 |

*Source:* Our own calculations based on a data from State Statistical Office

The HDI differs significantly between urban (0.796) and rural municipalities (0.765). This difference is particularly dramatic in respect of the "knowledge component", where the completed education index in the urban and rural municipalities reaches 0.890 and 0.810 respectively. The gross enrolment rate for all three education levels in the rural municipalities amounts to 62.36% and is significantly lower than the one in the urban municipalities - 78.52%.

## Human and Income Poverty

Income poverty is an important, but not the exclusive dimension of poverty. Poverty is a multidimensional and complex phenomenon, which in essence is about the lack of capabilities (opportunities) necessary for human development (education, health care and decent standard of living).

The level of poverty can be measured based on income and non-income measures. Simultaneous use of diverse poverty measures is a precondition for higher accuracy in estimating poverty level. Estimating poverty level is essential for three reasons:

1. To select an appropriate strategy for economic and human development.
2. To determine public consumption adequately.
3. To determine state intervention targets forwarded to alleviate poverty.

In this article Head count index, Poverty gap index and Foster-Greer-Thorbecke index are applied (all as measures of income poverty[4]) and Human poverty index - HPI (as a measure of multidimensional - non-income poverty). It should be taken into consideration that all these indexes are only estimations of real poverty. More precisely, they represent derivation of a particular sample, and therefore they are burdened with standard errors.

While the HDI gives information about the quality of human life from a conglomeration perspective, the HPI does so from the deprivation perspective. Nonetheless, both indexes go a step further from what the income indexes (GDP and the Head count index) can do (Sen and Anand, 1997). For the purposes of measuring the multidimensional poverty, we used HPI-2 index (Table 7.7):

**Table 7.7. HPI-2 for Macedonia (2004)**

| *Probability at birth of not surviving to age of 60 (times 100)* | *Adults lacking functional literacy skills (%)* | *Population below income poverty line (50% of median adjusted household disposable income* | *Rate of long-term unemployment (lasting 12 months or more) (%)* | *HPI-2 (%)* |
|---|---|---|---|---|
| 13.3 | 42.2 | 23.37 | 63.6 | 44.3 |

*Source:* Our own calculations based on data from State Statistical Office.

The HPI-2 shows that on the average 44.3% of the total population in the country is affected by various forms or deficits in respect of human poverty included in the index (Table 7.7). The high HPI-2 is mainly a result of the high rate of long-term unemployment[6] (63.6%), which is 6 to 10 times higher than in the highly developed countries, as well as of the high percentage of functionally illiterate people, which amounts to 42.2%, and is 2.5 to 7 times higher than in the highly-developed countries.

Income poverty measures are estimated on the basis of parameterized quadratic Lorenz curve in Table 7.8:

**Table 7.8. Income poverty in Macedonia (2004) – per household (relative method)**

| | |
|---|---|
| Head count index | 30.9% |
| Poverty gap index | 10.9% |
| Foster – Greer – Thorbecke index | 5.3% |

*Source:* Our own calculations based on data from State Statistical Office.

Head count index shows that 30.9% of the total population is under the poverty line, estimated on the level of 70% of the average household income (MKD 142,358.00 or about US$ 242 per month). Poverty gap index which counts 10.9% shows average proportional income shortage of the total population, or total income shortage by which the poor could reach the poverty line, whereas Foster-Greer-Thorbecke index counting 5.3% shows variations (changes) in the income share of the poorest households.

Elasticity of the poverty measures with regard to average household income and Gini index are showed in Table 7.9.

**Table 7.9. Elasticity of poverty measures with regard to average household income and Gini index**

| *Income poverty measures* | *Elasticity in respect to average income* | *Elasticity in respect to Gini index* |
|---|---|---|
| Head count index | -1,55664 | 1,03776 |
| Poverty gap index | -1,84245 | 2,89497 |
| Foster-Greer-Thorbecke index | -2,11914 | 4,74609 |

## Gender and Income Inequality

While HDI measures the average achievements, GDI (Gender development index) makes an adjustment of the average achievements and shows the inequality between women and men in terms of health (life expectancy), knowledge (literacy rate and gross enrollment rate) and the living standard (estimate of the generated income) (Table 7.10).[7]

The GDI[8] in Macedonia stands at 0.783 and puts Macedonia in the middle of the ranking list – close to Croatia (0.800) and Bulgaria (0.794). The GDI in the case of Macedonia shows moderate disparities, i.e. moderate gender inequality. The inequality is best reflected in the revenues generated by men and women (index 0.682) (Table 7.11), whereas the other forms of inequality are within the limits of what is a desired situation.[9]

According to the UNDP methodology, the GDI of 0.783 puts Macedonia in number 59 world ranking out of a total of 175 countries.

Perhaps one of the most important topics related to the economic development is the topic on (in)equality of income distribution, that is, the question on (un)equal economic growth distribution among the population. In this paper, as a measure of income inequality, Gini index is applied. Gini index avoids comparison with the average. It also avoids squaring and does not lose its sensitivity in income transfer from the rich to the poor. Therefore, we consider that this index, as inequality

**Table 7.10. Components required for the calculation of the GDI for Macedonia (2002)**

| Life expectancy (women) | Life expectancy (men) | Adult literacy rate (%) (women) | Adult literacy acy rate % (men) | Gross enrolment rate for all three education levels (women) | Gross enrolment rate for all three education levels (men) | Estimate of the generated income - women (PPP$)* | Estimate of the generated income - men (PPP$) | Share in the overall population (women) | Share in the overall population (men) |
|---|---|---|---|---|---|---|---|---|---|
| 1 | 2 | 3 | 4 | 5 | 6 | 7 | 8 | 9 | 10 |
| 75.21 | 70.68 | 92.95 | 97.10 | 68.71 | 72.09 | 4350 | 8600 | 0.497 | 0.503 |

*Source:* State Statistical Office and our own estimates.
*Note:* *Estimation based on the data from State Statistical Office.

**Table 7.11. GDI for Macedonia**

| Life expectancy index (men) | Life expectancy index (women) | Completed education index (men) | Completed education index (women) | Revenues index (men) | Revenues index (women) | Index of equally distributed life expectancy | Index of equally distributed education | Index of equally distributed revenue | GDI |
|---|---|---|---|---|---|---|---|---|---|
| 1 | 2 | 3 | 4 | 5 | 6 | 7 | 8 | 9 | 10 |
| 0.795 | 0.803 | 0.849 | 0.888 | 0.630 | 0.743 | 0.799 | 0.868 | 0.682 | 0.783 |

*Source:* Our own calculations made on the basis of data from State Statistical Office.

measure, provides best results. However, we in the interim take into consideration its failure. Namely, as a static measure it does not provide information on how the individual's position inside the income distribution is changed (a problem of almost all aggregate measures of inequality).

In Macedonia, Gini index of 22% in the beginning of the transition, increased to 35% in 2004 - showing moderate level of inequality. Such an increase in Gini index is mainly a result of private sector strengthening, where income dispersion is starting to extend in comparison to that of the public sector. Thus, Gini index in the private sector is 10% higher than that of the public sector. To illustrate, the difference between the highest-paid vorker in the private sector and the worker who is average-paid (in the same sector) is nearly 200%, while the difference in the public sector counts 60%.

Part of the population with the lowest wages mainly consists of women, manual workers, workers with primary or secondary education and workers who are employed in agriculture. Largest fraction of those profiles (about 50%) is employed in the private sector. It is worth mentioning that the head of a family (primary earner) is not usually low-paid (22%), it is a case with secondary earner (a wife or work capable children).

In the future, it is expected that the development of the private sector contributes to even higher level of inequality in the country.

According to the data in 2004,[10] annual average household income distribution can be showed as follow:

The histogram shows that the largest number of households in Macedonia realize annual income between MKD 70,945 and MKD 212,832 (from US$ 1,447 to US$ 4,343) confirming that the high percentage of households are under the poverty line (MKD 142,358.00, or about US$ 242 per month) estimated at 70% of the annual average household income. Participation of households that realize highest income is large and estimates 3%. In general, the annual income distribution of the households can be confirmed to be moderately unequal.

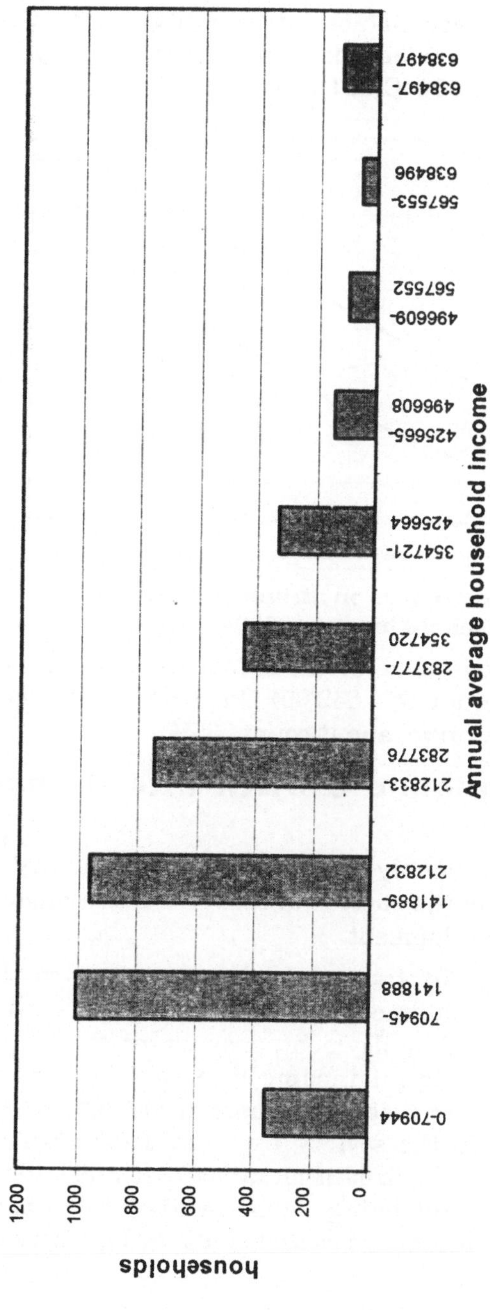


**Fig. 7.3. Annual average household income distribution (2004) in denars**

The area between diagonal (absolute equality) and Lorenz curve, visually shows income distribution inequality in Macedonia for the year 2004 (Chart 7.4)

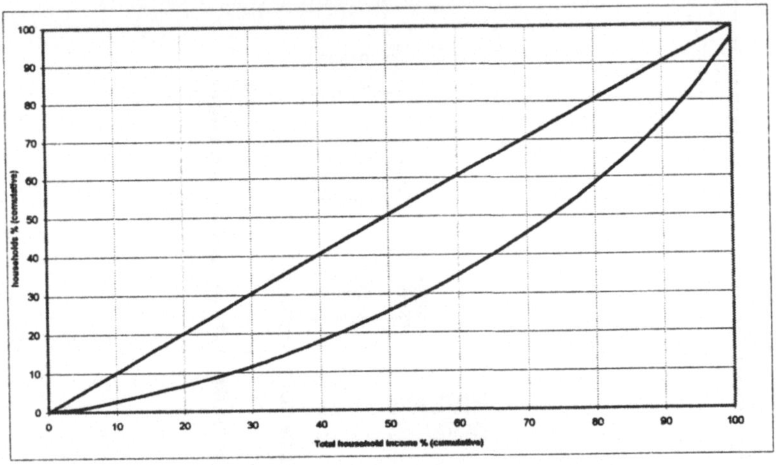


**Fig. 7.4. Lorence curve on an annual average household incomes (2004)**

Average annual household income in Macedonia estimates at MKD 237,263.60 (about US$ 4,842.00). Gini index is estimated on the basis of Lorenz curve, and it counts 35%.

## Converting Economic Growth into Human Development

The quality of life mainly is derived from the quality of economic growth, i.e. of the successful conversion of economic growth into human development.

If we consider "qualitative economic growth"- from the aspect of human development, as economic growth that generates higher level of employment, ensures security of the population and more equal income distribution among population, contributes to sustained human development and supports democracy in the society - we can conclude that economic growth in Macedonia does not satisfy essential criteria to be "qualitative", and being such, related to human development index, represents a component that depresses its value.

At this point we will shortly discuss the first three criteria that determine the quality of economic growth, which according to our opinion are of particular importance for Macedonia at this moment. They are: creating job positions, population security and income distribution.

With regard to creating job positions, it can be said that at least for the last 9-10 years Macedonia is a classical example of a country with so-called "jobless growth", i.e. country employment stagnates regardless of the positive economic growth rates. As employment rates do not "follow" economic growth rates, it appears that the first criterion of "qualitative economic growth", from the aspect of human development, is not fulfilled (See, Fig. 7.5).

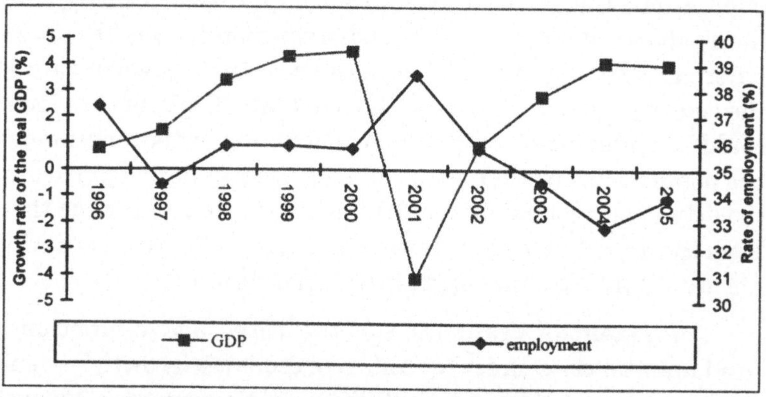


**Fig. 7.5. Growth of the real GDP and the rate of employment**

*Source:* State Statistical Office and Ministry of finance of Republic of Macedonia

With regard to the second criterion, security of the population (health security, personal security, political security and etc.) and in this context economic security, the results are partial. Though on one side health, political and personal security is on a satisfying level the same does not stand for the economic security of the citizens. Namely, if we consider, as a criterion for economic security of the citizens, their available (sufficient) income that could satisfy their basic needs, then we can conclude that economic security of the citizens of Macedonia is extremely alarming. Fundamental indicators for this condition are surveys carried out by State Statistical Office, aimed to measuring "subjective poverty" of the households. According to these surveys for the year 2004, only 3.2% of households reported that their monthly income wholly satisfies their needs, 27.4% reported that are mainly satisfied, and 69.4% of households reported that their monthly incomes mainly do not satisfy, or do not satisfy their needs at all. Due to the previous, we can conclude that the second criterion for "qualitative economic growth", from the aspect of human development, is in part fulfilled.

With regard to the third criterion for "qualitative economic growth", i.e. equal income (growth) distribution among the population, the matters are more complicated. Namely, Gini index shows moderate - income distribution inequality (35%), whereas income poverty measures show high poverty rate in the country (30,9%). However, the moderate height of Gini index could be understood as equality in distribution of poverty among the population. In other words, among the population particular poverty level instead of "wealth" is distributed. It means that, from the aspect of human development, even the third criterion for "qualitative economic growth" is not fulfilled.

The extent of realization of the three above-mentioned fundamental criteria for "qualitative economic growth" reveals the common statement that converting economic growth into human development in Macedonia is not at a satisfactory level.

## Concluding Remarks

Human development index (HDI) for Macedonia, as a measure of the average achievements in the field of human development, counts for 0793, which ranks Macedonia number 60 out of a total of 175 countries in the World. It significantly differs from the average HDI of the highly developed countries - 0.911, which can not be said about the Central and Eastern European countries.

The HDI at the level of aggregated urban and rural municipalities shows significant disparities between different types of municipalities. The HDI for urban municipalities reaches 0.796, while the HDI for rural municipalities reaches only 0.765. The most significant difference is related to the "knowledge component".

HDI height, in the case of Macedonia, does not confirm the hypothesis that there is no automatic link between the growth of the real income per capita (economic growth) and the growth of the HDI (human development). The insignificant disparities between these two components are mainly result of the inequality in the income distribution among the population.

The Gender Development Index (GDI) shows only minimal deviation between genders (0783) and the inequality is most pronounced in income component (revenues generated by men and women). It puts Macedonia in the middle of the World ranking list.

Human Poverty Index (HPI-2) shows that on the average 44.3% of the total population in the country are affected by various forms or deficits included in this measure. The high HPI-2 is mainly a result of the high rate of long-term unemployment which is one of the highest in Europe - 63.6%.

The extent of realization of the three major criteria for "qualitative economic growth" (creating job positions, population security and income distribution) reveals the conclusion that converting economic growth into human development in Macedonia is not at a satisfactory level. Therefore, we emphasize that the economic growth should ultimately result in improved quality of human life. People should be given opportunities and chances to participate in the economic activities, while at the same time enjoying the effects of the increased GDP. The quality of human life should be emphasized along with the quantitative aspects of the economic growth.

Our general conclusion is that thus far, the transition has failed to produce the expected positive effects. On the contrary, the economic efficiency of the system is weakening and the social security and the quality of life of the population is still deteriorating. The creators of macroeconomic policy should have in mind that the macroeconomic stability is not the ultimate goal of the development process, but a means to achieve the basic goal – quality of life of the people. Therefore, the macroeconomic policy of stabilization should not be understood as a synonym of the macroeconomic and human development policy.

## NOTES

1. Revised and extended version. Original title: "Measuring quality of life in Macedonia - using human development indicators", Proceedings of Rijeka Faculty of Economics - *Journal of Economics and Business*, Vol. 24, No 2, 2006, pp. 257-272.
2. The first component - *life expectancy* is measured by the expected length of the persons from birth. The second component - *the achieved education* is measured through two variables : *Adult literacy rate* and the *gross enrollment rate for all three educational levels* (primary, secondary and tertiary), i.e. through the synthetic measurement of the achieved education (E):

   $E = a_1 P + a_2 G$;

   where $a_1 = \frac{2}{3}$ and $a_2 = \frac{1}{3}$

   are parameters for the literacy of the adults (P) and the enrolled students at all three levels of education (G), respectively. The third component - *the living standard* is measured by the real gross domestic product (GDP) per capita (PPP$).

   Human development index estimation is carried out in three consecutive phases, where: *The first phase* reveals which of the three above-mentioned components - live expectancy at birth ($X_1$), achieved education ($X_2$) and real GDP per capita ($X_3$) - is scarce in the relevant country. For estimation of component indexes the following equitation is used:

   $$I_{ij} = \frac{(X_{ij} - \min_j X_{ij}}{(\max_j X_{ij} - \min_j X_{ij})}$$

   where index $I_{ij}$ is scare indicator for j – country in regard to i - variable, whereas, maximal and minimal component values are standardized and fixed for all countries:

   Life expectancy at birth 25 – 85 years
   Adult literacy rate 0% - 100%
   Combined primary, secondary and tertiary
   gross enrolment ratio (%) 0 - 100

GDP per capita (PPP) US$ 100 – US$ 40,000

Scarcity of a particular component sets the country in rang 0-1. Third component index estimation – income, is carried out with the equitation:

$$W(y) = \frac{\log y - \log y_{min}}{\log y_{max} - \log y_{min}}.$$

*In the second phase* scarcity index ($I_j$) is determined:

$$I_j = \sum_{i=1}^{3} I_{ij}.$$

*In the third phase* Human Development Index (HDI) is counted.

3. According to the UNDP methodology, countries are divided in three groups depending on the height of the human development index, as follows:
   - Countries with high human development - over 0.800 HDI,
   - Countries with middle human development - from 0.500 to 0.800 HDI,
   - Countries with low human development - below 0.500 HDI.
4. Head Count Index shows the percentage of population whose income or consumption per capita is under the poverty line:

$$H = \frac{q}{n},$$

where H = Head count index; q = number of people who are considered poor; n = total population. Poverty Gap Index measures the total deficit of all poor people:

$$PG = \frac{1}{n}\sum_{i=1}^{q}\left[\frac{z - y_i}{z}\right] \text{ for } y_1,\ldots\ldots, y_q < z < y_{q+1},\ldots\ldots, y_n,$$

where: PG = Poverty gap index; z = poverty line; $y_i$ = income of the i-individual. Foster-Greer-Thorbecke Index belongs to the group of so-called General Class Poverty Measures, i.e. poverty measures from P – alpha class. In its original shape, it is in fact Poverty Severity Index:

$$P2 = \frac{1}{n}\sum_{i=1}^{q}\left[\frac{z - y_i}{z}\right]^2$$

where $P_2$ = Poverty Severity Index.

5. HPI-1 is a measure of human poverty for developing countries, while HPI-2 for selected OECD countries. For more details see any Human Development Report.

   The equation for calculating the HPI-2 is as follows:

$$HPI-2 = \left[\frac{1}{4}\left(P_1^{\alpha} + P_2^{\alpha} + P_3^{\alpha} + P_4^{\alpha}\right)\right]^{\frac{1}{\alpha}}$$

Where:

$P_1$ = probability at birth of not surviving to age of 60 (times 100);

$P_2$ = adults lacking functional literacy skills;

$P_3$ = population below income poverty line (50% of median adjusted household disposable income;

$P_4$ = rate of long-term unemployment (lasting 12 months or more);

$a$ = 3

6. Data based on Labour Force Survey (2005) and our estimations.
7. For more see: Sen, Amartya and Anand, Sudhir (1995): *Gender Inequality in Human Development: Theories and Measurement*, UNDP Working Papers.
8. The calculation involves three steps. First, female and male indices (Table 7.11-columns 1 to 6) in each dimension (see: Table 7.10) are calculated using equation:

$$Dimension\ index = \frac{actual\ value - min\ value}{max\ alue - min\ value}.$$

Second, the female and male indices in each dimension are combined in a way that penalizes differences in achievements between men and women. The resulting index, referred to as the equally distributed index, is calculated according to this equation:

Equally distributed index = {[female population share (female index$^{1-\epsilon}$)] + [male population share (male index$^{1-\epsilon}$)]}$^{1/1-\epsilon}$

$\varepsilon$ measures the aversion to inequality. In the GDI $\varepsilon = 2$. Thus, the general equation becomes:

Equally distributed index = {[female population share (female index$^{-1}$)] + [male population share (male index$^{-1}$)]}$^{1/1-\epsilon}$

which gives the harmonic mean of the female and male indices (Table 11 - columns 7 to 9). Third, GDI is calculated by combining the three equally distributed indices in an unweighted average (Table 11 - column 10).
9. If there were gender equality the GDI would equal the HDI.
10. Estimations are done on the basis of a survey on household consumption (2005), State Statistical Office, sample: 5040 households (1% of the total number of households in the country).

## REFERENCES

Chesterman, Simon (2005), "State-Building and Human Development", *Background Paper, UNDP.*

Consumption of households in the Republic of Macedonia (2005), State Statistical Office.

Dikhanov, Yuri (2005), "Trends in Global Income Distribution 1970-2015", *Occasional Paper, UNDP.*

Eftimoski, Dimitar (2003), *Economics of Development*, Institute of Economics - Skopje, Skopje.

Etimoski, Dimitar (2004), "Human Development, Inequality and Poverty in Republic of Macedonia in Transition Period", *Global Scholarly Publications*, New York.

Eftimoski, Dimitar (2006), "Conventional Poverty Measures and Measures Derived from Parameterized Lorenz Curve", *Economic Development, Journal of the Institute of Economics - Skopje.*

Eftimoski, Dimitar (2006), "Measuring Quality of Life in Macedonia - Using Human Development Indicators", *Proceedings of Rijeka Faculty of Economics - Journal of Economics and Business*, Vol. 24, No. 2.

Fuentes, Ricardo (2005), "Poverty, Pro-Poor Growth and Simulated Inequality Reduction", *Occasional Paper, UNDP.*

Fukuda Parr, Raworth, Shiva Kumar (2002), "Using the HDI for Policy Analysis", *Oxford Training Course - 2002*, mimeo, UNDP and QEH.

Goodman, Alissa (2005), "The Links between Income Distribution and Poverty Reduction in Britain", *Occasional Paper, UNDP.*

Human Development Report, (1990-2005), UNDP.

Jahan, Selim (2002), "Measuring Human Development: Evolution of the Human Development Index", *Oxford Training Course - 2002*, mimeo, UNDP and QEH.

Kabeer, Naila (2005), "Gender Equality and Human Development: The Instrumental Rationale", *Occasional Paper, UNDP.*

Keith Griffin and Terry McKinley (1994), *Implementing a Human Development Strategy*, MACMILLAN.

National Human Development Reports (NHDR) - Macedonia (1998; 2001; 2004), UNDP.

Sen, Amartya and Anand, Sudhir (1994), "Sustainable Human Development: Concepts and Priorities", *UNDP Working Papers.*

·Sen, Amartya and Anand, Sudhir (1995), "Gender Inequality in Human Development: Theories and Measurement", *UNDP Working Papers.*

Sen, Amartya and Anand, Sudhir (1997), "Inequality, Unemployment and Contemporary Europe", *International Labor Review*, Vol. 36, No.2.

Sen, Amartya and Anand, Sudhir. (1999), "The Income Component in HDI - Alternative formulation", *UNDP Working Papers.*

Sen, Amartya (2000), "A Decade of Human Development", *Journal of Human Development*, Vol. 1, No.1, February.

Stewart, Frances (2002), "Human Development as an alternative development paradigm", *Oxford Training Course - 2002*, mimeo, UNDP and QEH.

# 8

# Social Capital Formation in Rural, Urban and Suburban Communities

*David L. Debertin*
*Stephan J. Goetz*

## ABSTRACT

In this paper, we define social capital from different perspectives and show how it is linked to the concept of a community. Based on these definitions, a conceptual framework for analyzing and measuring social capital and its indicators is developed. A typology for analyzing social capital is then created based on different types of communities. The characteristics of three prototype communities – a small rural community, a modern city suburb, and a community located in the core of a central city are outlined. For each prototype community, social capital formation strategies and indicators suggesting evidence of social capital are identified. Implications for future research efforts dealing with social capital are discussed.

**Key Words** : social capital, community, rural community, suburban community.

## Introduction

Social capital refers to the "... stocks of social trust, norms and networks that people can draw upon in order to solve common problems" (Sirianni and Friedland 1997). These networks involve activities of "civic engagement" such as volunteerism and participation in neighbor-hood associations, service clubs and charitable groups. A rapidly expanding literature exists on social capital and its importance to rural and urban areas.[1] In both rural and urban areas, social capital refers to the institutions and mechanisms where by residents relate to and interact with each other to solve problems for the common good (Ostrom, 1994).

In this article, we first provide basic definitions of social capital and how social capital is linked to the concept of a community. We then develop a conceptual framework for analyzing and measuring social capital and its indicators. A typology for analyzing social capital is created based on different types of communities. Social capital formation strategies and social capital indicators are examined for rural, suburban and urban communities.

## Communities and Social Capital: Definitions

Coleman (1988) used the term social capital to refer to all human relationships and described social capital in functional terms as "the value of those aspects of social structure to actors as resourc-es that they can use to achieve their interests" (S101). Schmid and Robison (1995, p.58) see social capital as embodying obligations, expectations and trustworthiness of structures, information channels, norms and sanctions. The concept of a community and social capital are intertwined.[2] In rural development, community frequently describes residents of a small town and its surrounding area (Salamon, 1995).[3] Presumably, those living in such a rural community share many common interests and values and thus the geographic and non-geographic definitions may overlap in rural areas (Broom and Selznik 1963, p.19-22).

Social capital formation is the development of networks in which community residents can identify problems, share information, and implement strategies designed to solve problems for the benefit of all. This invariably leads to the questions of how social capital is formed, how its formation can be fostered, and what makes some rural and urban communities have high social capital and others have low social capital (Fine, 2001, Mesch 1996; Newton 1997).

The conceptual framework we propose here examines the social ends ($y$) to be achieved in a given setting as being a function of social capital stocks ($s$) and other exogenous factors ($w$) that detract from or facilitate accomplishing the social ends. That is,

$$y = f(s, w)$$

Here the function $f$ captures the efficiency with which $s$ and $w$ are translated into $y$[4]. A key challenge is to separate the definition of social capital from the indicators that are used to measure it. Unlike physical capital (infrastructure) and human capital (education), social capital is difficult to measure. At best, we can propose proxy indicators for social capital, and discuss how the indicators vary with the social ends to be achieved and with the definition or size of the community being analyzed.

A *social capital indicator* is any quantifiable measure thought to vary with levels of social capital. Commonly used social capital indicators include membership in civic (Jaycees, Lion's Club, Kiwanis, Chambers of Commerce, Neighborhood Watch Associations) and fraternal (Fraternal Order of Police) organizations; measures of citizen participation including voter turnout rates; various measures of volunteerism; participation in parent-teacher associations, religious groups and the like (see the discussion on social capital measures in Putnam 1993*a*). The President's Council on Sustainable Development (1995) highlights the importance of civic engagement as a social capital indicator.

Social capital can be analyzed in a number of dimensions, including in terms of the ends to be achieved by social action, the type of community under consideration, and the degree of demographic or socioeconomic diversity that exists among the relevant actors within the community. This is illustrated using a cube in Figure 1, with appropriate $x$-, $y$- and $z$-axes. Distinguishing among these dimensions is important because the type, meaning and efficacy of social capital depend on the area of the cube in which one is conducting the analysis.

For example, a community ($x$) might be a nuclear family, an urban or a suburban neighborhood, a rural community, a region within a state, a national region, or even an entire nation.

The type and effectiveness of social capital to achieve social ends will vary depending on the size of the community. Similarly, the degree of community heterogeneity ($z$) can vary from completely homogenous to highly diverse, as measured by race, class, income or occupation. Analyses of the effect of ethnic heterogeneity are relevant within families (mixed-race marriages), neighborhoods or nations.

However, we do not suggest that homogeneity, *per se*, promotes social capital within a community such as a neighborhood. There are many examples of sub-divisions with homogeneous populations and little social capital, on the one hand, and lifestyle enclaves such as Bohemian urban neighborhoods with diverse populations that exhibit a high degree of social capital, on the other hand. Even so, we believe it is useful to include heterogeneity of the population as one of the factors driving social capital formation, if only to highlight this dimension and the need for research to determine why social capital forms in some environments but not in others, with populations that otherwise of the same degree of heterogeneity.

The ends achieved by social action ($y$) may range from social and economic goals (reducing poverty, protec-ting home property values and preventing teenage pregnancy) to moral ends (desegregation or eliminating racist complicity). To illustrate, reduction of poverty is a social goal that can be addressed at the level of individual families, urban and suburban neighborhoods, rural communities, or even entire nations. In contrast, the goals of reducing teenage pregnancy and dropping out of high school are perhaps best dealt with using capital at the level of families or through neighborhood effects (Katz, 1992, Goetz, 1993, also see Clinton's 1996. It is questionable whether national-level or federal policies can effectively reduce teenage pregnancy, while state-wide minimum age school attendance legislation may, at least superficially, reduce high school dropout rates.

## Social Ends and Indicators of Social Capital in Different Communities

Different units of analysis or types of communities in which social capital has been studied, examples of social goals to be achieved in each community, and illustrative examples of indicators of both produc-tive and unproductive forms of social capital under each unit, as they affect the achievement of the social goal. The fact that a particular indicator of social capital appears in only one cell of the table does not mean that the type of social capital arises only in that community. For example, Kiwanis club membership is listed under the City/Urban Core because it is one indicator of social capital potentially relevant to the problem of preventing urban decay (Hornburg 1998). Obviously, Kiwanis club members can also live in suburban neighborhoods, but membership in such a club in all likelihood has limited relevance to the problem of protecting property values. While we also briefly review social capital issues related to nations or regions, our main interest is in discussing city and suburban neigh-borhoods and rural communities, as they are most germane to questions of housing policy and real estate values.[5]

## *Rural Communities*

Many rural communities have experienced economic decline–often in tandem with the inner core of urban areas—as a result of agricultural industrialization and globalization, combined with increasing skills demands associated with technological change (Goetz and Debertin, 1995). As a result, these com-munities face challenges such as creating employment opportunities, stimulating growth and development, and reversing the outflow of people. Net population losses, of course, have implications for housing values. Recently, Americans have started to move into rural areas in search of amenities not available in suburbs.

Rural communities are widely believed to have high levels of social capital, which can some-times work to prevent undesired changes from occurring. Beggs, *et al.* (1996, p.316) conclude that "personal networks of non-metropolitan residents contain a larger proportion of long-term relations . . . tend to be more multiplex . . . contain higher proportions of kin, neighbors, and kin-neighbors . . . are significantly smaller and . . . dense . . . [and]

exhibit significantly higher religious heterogeneity" than do networks involving metropolitan residents. Petee and Kowalski (1993, p.88) find that rural areas with greater residential mobility, racial heterogeneity and shares of single-parent households have higher violent crime rates, suggesting that stocks of social capital (as reflected in violent crime rates) are not uniformly high in all rural communities.

If the rural community is sufficiently small, nearly the entire community may function as a single social capital network (see the discussion on community network building in Ayres 1996, pp.21-36). This feature makes rural communities inherently different from urban settings, where housing is usually segregated by income, education, race and type of employ-ment (Massey and Denton 1993). Green (1996, p.7) maintains that the close knit nature of rural communities may impede economic development, because of a reluctance of residents to form links with new groups both within and outside their communities. With economic development comes change: For example, a new industrial plant locating in a rural community may bring with it workers of lower or higher incomes and different cultures (Debertin 1993; Debertin, Infanger and Goetz 1991).[6] Thus, depending on one's perspective, unproductive forms of social capital in rural areas include a general apprehension of those who did not grow up within the same commu-nity. In addition, Flora (1997, p.115) derides "[b]edroom communities [and] rural communities that become a low-rent haven for jobless urbanites" as lacking interaction among residents that results from a lack of trust.

## Suburban (and Urban) Neighborhoods

"Neighborhoods were created within cities to enable people to live in a small er-more community-like-environment" (Moore 1991, p.355), and home buyers search for homes in locales described by others as "good" or "nice" neighborhoods. Weiss (1989, p.32) describes the changing motivations of individuals over their life cycles in selecting neighborhoods:

To those in their 20s, the prime consideration in selecting a home is its suitability for raising young children. Those in their 30s are more concerned with the quality of local schools. From 40 to 55, ease in commuting is important. Older folks want safety from crime.

Residents living in a particular location within an urban or suburban community often form geography-based linkages with their neighbors.

In urban communities, the location within the city often reveals something about a resident's economic and social class, the type of occupation (white collar, blue collar) and perhaps even religion (residents following a particular religion perhaps locating in neighborhoods surrounding a parochial school). In a typical urban setting, highly-paid business executives, success-ful entrepreneurs, doctors and lawyers likely live in certain subdivisions or neighborhoods—those with comparatively high-priced homes and well-manicured lawns. Those lower on the economic and social ladder, say teachers, or those engaged in "medium-income" white-collar occupations, live in less up-scale, though still "good" neighborhoods.

Traditional blue-collar neighborhoods are inhabited by individuals with high school educations or less, again with net-works of shared interests and values. These also can be "good" neighborhoods with high social capital (Broom and Selznik 1963, p.190-201).[7] Indeed, some of the best examples of social capital networks evolve from situations whereby residents of modest means work to improve the community for the common good. Examples of these include small-scale urban gardens located in areas adjacent to public housing projects, neighborhood beatification projects that involve planting street trees, efforts to build bicycle and hiking paths, and efforts aimed at converting a vacant lot into a park or playground area.

Since housing is for most individuals the single largest expenditure they incur, home owners have a strong interest in preserving if not increasing the value of their property over time (Laird 1996, p.24; also Ray, Halseth and Johnson 1997, p.96). One of the most influential writers on neighborhood dynamics is Grigsby (1963), whose contributions are enumerated in Megbolugbe, Hoek-Smit and Linneman (1996). Grigsby argued that "[t]he best and newest housing is built for higher-income groups but eventually passes down to successively lower-income groups until it reaches groups so low that they cannot afford to cover maintenance costs. Once a neighborhood reaches this stage . . . public intervention is necessary either to restore the neighborhood to health or to clear it for reinvestment in new structures" (Megbolugbe, *et al.*, 1996, p.1782). Grigsby's dynamic model for analyzing neighborhood change can be summarized as (*ibid.* 1785, quoting Grigsby *et al.,* 1987, p.33):

Changes in social and economic variables (1) cause households acting directly or through a system of housing suppliers and market intermediaries (2) to make different decisions regarding level of maintenance, upgrading, conver-sion, whether to move, new construction, boarding-up, and demolition (3), producing changes in dwelling and neighborhood characteristics. A key challenge for researchers is to discern whether and how different levels and types of social capital can attenuate any negative consequences of such changes, to achieve social goals.

Threats to property value in a neighborhood community include externally-imposed locally undesirable land-uses (LULUs), pressure to develop land, physical deterioration, change in population, rising crime rates, and different exogenous influences have different impacts in different types of communities (Mesch 1996, p.346; O'Looney 1993; Freeman, Grogger and Sonstelie 1996; Megbolugbe *et al.,* 1996, p.1790). Whether or not social capital is one means of protecting property over time is a researchable question. Hirschfield and Bowers find that "[t]he more that an area that is at a disadvantage economically pulls together as a community [i.e., is socially cohesive], the greater its capacity to combat crime."

Racial or ethnic heterogeneity is an important factor often mentioned in the literature as influencing social capital levels. While examples exist of how racial diversity can lead to greater stability and resiliency to external forces within a community (as discussed at the beginning of this section), a sizeable literature suggests that racial heterogeneity is associated with low levels of social capital. For example, Hirschfield and Bowers (1997, p.1290) find that "juvenile disturbances, single-parent families and social heterogeneity" and "high levels of migration and ethnic heterogeneity," as measures of a lack of social cohesion, are associated significantly with higher crime rates in low-income areas (and thus reduced property values). Neighborhood associations are key institutions for coping with external or internal influences on property value. According to Mesch (1996, pp.346-7), neighborhood associations

> . . . are civic associations whose goals are to maintain and improve the quality of neighborhood life and to protect the common and social interests. . . . Neighborhoods that lack organizations are slower and ineffective in their responses [to external threats . . . while those] with preexisting organizations respond faster and are more effective in rejecting or making important changes to development plans.

Factors such as neighborhood stability (percent of residents who had lived in the neighborhood for more than five years) and home ownership (percent of residents who own rather rent their homes) were associated with a greater likelihood that a neighborhood association existed, according to one study in Columbus, Ohio. Also, threats of undesired change as well as the ability of residents to bear the costs of involvement in the association affected the probability that an association was formed (Mesch 1996, p. 363). The same study suggests that neighborhood associations in wealthier areas are less active, largely because they face fewer environmental threats (356).

A central contribution of Mesch's work is to highlight the importance of separating the formation of a neighborhood association from the actions of a neighborhood association once it is in existence. Furthermore, it points to certain social capital-type proxies that can be used to predict whether or not an association is likely to form in a neighborhood to preserve property values. Portney and Berry (1997, p. 642) argue that

If sense of community is an essential building block for creating a participatory democracy and transforming the way people behave in the political process, then the cities with citywide systems of neighborhood associations have a valuable resource. These neighborhood associations are more effective than other types of organizations *in nurturing feelings of identity and shared purpose with one's neighbors*. [*emphasis added*]

In terms of participation in civic activities by different racial groups, Portney and Berry (643) conclude that "[a] though neighborhood associations do not increase the overall number of people who participate in city politics, minority participation rates tend to increase as their percentage of the neighborhood population increases."

Another important institution directly or indirectly governing property values in a neighborhood is that of covenants or deed restrictions. These are "credible commitment(s) by unknown future neighbors to utilize their own property within the defined bounds, thereby reducing the uncertainty or riskiness of future externality effects" on property values (Hughes and Turnbull 1996, p.160). Whether these covenants are associated with high or low levels of social capital is an interesting but unanswered question. Less uncertainty exists about the effect of such covenants on property values. In a study of 37 Baton Rouge, LA neighborhoods and 1,314 property transactions, Hughes *et al.* find "housing price capitalization patterns consistent with our risk abatement theory, with the value of a given set of deed restrictions diminishing as the neighborhood matures" (1996, p. 171).

Gated communities represent an increasingly common type of development in suburban areas, particularly in the Sun Belt.[8] Wilson (1997, p. 747) dismisses gated communities as "manifestations of an increasingly polarized global economy" and concludes that these communities reflect a negative form of social capital in suburban societies. Lang and Danielson (1997) argue, however, that social capital relationships within gated communities are far more complex. Brick walls, high fences, iron gates and other natural or man-made architectural barriers seem designed to provide privacy from those living outside the gated community and at the same time encourage social interaction and cohesiveness for those within. They further suggest that while residents believe the gates and fences offer protection from vandalism and burglaries, the barriers are usually more psychological than real (1997, p. 876). Within a gated community, residents may form cohesive social bonds for ensuring that property is properly maintained and the welfare of residents is protected. Third parties are often used to resolve disputes among neighbors (Lang and Danielson, 1997, p.873).

### City/Urban Core

If protecting property values and avoiding change are the major social ends to be achieved in suburban and urban neighborhoods, effecting change to revitalize downtown areas has become the main objective of many American metropolitan areas. While much of the discussion in the previous sections also pertains to residential neighborhoods in the core areas of cities, the twin extremes of urban flight and gentrification separate urban from suburban neighborhood problems. Flora (1997, p. 115) also characterizes communities "undergoing gentrification" as lacking trust and interaction among residents.

Of equal importance for the analytical framework we are proposing here, the types of social capital that arise or fail to arise to meet social ends also differ in city as opposed to suburban neighborhoods. Church attendance and Kiwanis Club, Chamber of Commerce or bowling league membership rates may be important indicators of social capital that can be mobilized to revitalize a downtown area, but these indicators most likely reveal little about the ability of a suburban neighborhood to prevent a LULU.

### Region and Nation

Only relatively limited work has been carried out on the effect of social capital on the economic growth of regions and countries. The seminal works are those of Putnam (1993*b*, 1995, 2000) and Helliwell and Putnam (1995). (See a critique of Putnam's 2000 book in Wills.) More recently, Knack and Keefer (1997) used data on trust and adherence to civic norms from the World Values Surveys found both of these indicators of social capital to significantly affect economic growth between 1980 and 1992 in a sample of 29 countries. Greater income equality and ethnic homogeneity were found to be associated with higher levels of trust and civic cooperation in a society, while the percent of students pursuing a law degree is associated only with lower levels of trust in a society (1282). This confirms their argument (1252) that much of the economic backwardness in the world can be explained by the lack of mutual confidence . . . [since] [i]ndividuals in higher-trust societies spend less to protect themselves from being exploited in economic transactions. Written contracts are less likely to be needed, and they do not have to specify every possible contingency. Litigation may be less frequent.

Knack and Keefer (1284) also found that ". . . horizontal networks!as measured by membership in groups [as proposed by Putnam, 1995] are unrelated to trust and civic norms . . . and to economic performance; Kenworthy (1997) arrives at a similar conclusion. Thus, it is clearly critical to be sensitive to the different types of social capital, and ways of measuring them, in these kinds of analyses.

Unproductive social capital has also been found at the level of nations or regions. Rubio (1997) develops the thesis that social capital which fosters criminal activity leads eventually to economic decline. Using data from Columbia, he fails to reject his thesis. Thus, clearly not all forms of social capital lead to higher general economic and social well-being.

DeFilippis, (2001) Offers an excellent critique of Putnam's work. He compares Putnam's work with that of French sociologist Pierre Bourdeiu (1985) and notes that Putnam usually measures social capital simply by counting individuals who participate. Putnam believes that social capital is a resource that people or groups either possess or do not possess, and is not something that can be measured on a continuous scale. Putnam emphasizes the positive not the negative components.

In contrast, Bourdeiu sees social capital as embedded in social relationships and is realized by individuals. He is much less inclined to measure social capital simply by counting members of societies and community organizations. He is much more concerned about the phenomenon of negative social capital and notes that organizations such as terrorist "cells," the Mafia and the KKK have social capital of sorts, but this is not positive.

## Social Capital in Three Idealized Prototype Communities

In this section we describe and analyze how the physical layout of the community may affect social capital formation in three idealized prototype communities The top panel represents characteristic features often found in small, self-contained rural communities common to rural areas in the Midwest that were settled in the late 19th and early 20th Century. The middle panel illustrates features likely to be found in late 20th century suburban developments. The bottom panel shows patterns of city-core urban residential develop-ment.

### *The Small Rural Community*

This park or town square provides a focus for community-wide gatherings and a place for children to play while parents shop as well as a place for residents to meet and interact. Businesses are within easy walking distances from most homes. Homes built prior to World War II—frequently one and one-half (Cape Cod or Craftsman style) two-story (Colonial) structures—are located on narrow lots that line the streets directly adjacent to the business and the central square. Homes constructed after World War II–frequently of a single-story or "ranch" style–occupy wider lots slightly more distant from the town square, but still within walking distance.

The street pattern–even within these newer sections–is still a grid, with all streets carrying similar amounts of traffic. The light industrial area–usually involving agriculturally-related business-es—is located adjacent to the railroad tracks on the outskirts. Other agriculturally-related businesses often line the streets leading to this light industrial area. Churches often are located on corner lots of blocks, with residential housing on the same block. Modern school buildings constructed in the 60s and 70s (not shown in this illustra-tion) are frequently single-story structures built on expansive locations at a half mile or more away from the town itself.

The geographic and non-geographic social capital networks overlap in small rural communities such as the one illustrated here, making it difficult to distinguish between the two. In fact, usually only a single social capital network exists in small, isolated rural communities, which is generally considered to be a strength or asset of such communities (Green 1996, p. 7). Many rural communities similar to the one illustrated here were originally planned with a design intent on fostering social interaction. For example, county seat towns in many states were built around a plan in which the county courthouse is positioned on a central square (replacing the park), with major shopping areas facing the square on each side.

Rural communities of the size illustrated here are usually not of sufficient size to attract discount stores and other threats to main street businesses, and these main street businesses remain at the center of community social interaction. Often, the major state highway route was along main street – the planning idea being to attract travelers to main street businesses. As rural highways have been refur-bished in the past few decades, they are often rerouted away from main street routes and businesses in an effort to speed traffic flow (the "by-pass" phenomenon), which can lead to business closings along main street.

Because homes of varying values in rural communities are often located in comparatively close proximity to each other, the potential for social interaction among people of (moderately) varying income classes increases, as does *trust* across economic lines. As communities become larger, homes in a similar price range are more likely grouped together, increasing the likelihood that individuals of similar economic backgrounds will interact, but potentially reducing interaction among those of different income levels (Massey and Denton 1993).

A Newsweek article reported that some small rural towns near Chicago and elsewhere were being invaded by wealthy of the city (Adler *et al.* 1995). Wealthy city residents find relatively in-expensive land and houses, clean air and a commute that is acceptably short; they buy land or old housing cheaply in the rural community, and often demolish existing homes to build three-story neo-Victorian homes in their place. This forever changes the character of the rural community, and not always for the better, at least according to long-time residents.

## Suburban Development

The middle panel is an idealized map for a prototype *late 20th century suburban development*. Lewis (1995, p. 43) points out that many of these communities are carved out of farmland, leap-frogging over rural space, and thus may not be directly adjacent to an urbanized area. The majority of homes in the idealized map are on streets that lead no where, that is, courts or cul-de-sacs.[9] Developers often take advantage of land which contains a natural feature such as a small lake or stream, as illustrated here. Often the more expensive homes are located in the most secluded areas of the subdivision (and backing to the water, in this case), and homes facing streets that feed traffic to the four-lane feeder arteries are priced lower than houses located on the dead-end streets. A mandatory fee (subdivision tax) pays for upkeep of the subdivision club, and common areas providing a pool and other recreational facilities for local residents. The subdivision club serves a function analogous to the town square in the small rural community as discussed in detail below.

The city may have required the developer to donate land for an adjacent public park for use by residents and non-residents alike, but these areas are often not centrally located and do not constitute a social capital focus for the subdivision itself. In subdivisions developed in the 60s and 70s, homes consisted of a mixture of single and multi-story designs, but in recent years, as land become more and more expensive, multi-story designs that make more efficient use of land increasingly predominate.

- An elementary and middle school is perhaps three miles away, but the new high school might be reached only after a 20 minute drive.

The presence or absence of social capital is indicated in very different ways in the typical suburban community. Subdivisions in communities are frequently designed to contain homes only within a relatively narrow price range. Because of this, in many of these communities, the subdivision in which an individual lives thus becomes a clear indication of social and economic status and perhaps even political influence within the community. This, in turn, is reflected in public services provided at taxpayer expense. Homes in the "right" neighborhoods are served by the "best" schools, although the direction of causality is an issue here.

In upscale suburban neighborhoods much evidence exists of "shared values" among residents. These shared behaviors do not necessarily suggest that residents have a great deal of contact with their neighbors.[10] Certain suburban communities across the country have been criticized for being impersonal, with even those living in close proximity often not knowing each other well;[11] automatic garage door openers can significantly reduce interaction among neighbors, for example. But, since the value of one's home in such a setting is often heavily influenced by the value of neighboring homes, those living in such communities usually pay close attention to neighbors' actions that may even slightly depress neighborhood property values. This has led to elaborate covenants and deed restrictions on individual property owners with respect to what is (or is not) permitted, including restrictions prohibiting detached garages or other detatched structures, regulations involving the use of residential streets for overnight parking of vehicles, and others. These are similar to, but in some respects more subtle than, the types of deed restrictions analyzed by Hughes and Turnbull (1996) in Louisiana. Modern suburbs have evolved a great deal from the idealized prototype. Traditionally, suburbs such as the one illustrated served as bedroom communities for those working and often shopping in the city core. Schools, churches, grocery stores and gas stations quickly were built near the suburban homes. In the 60s, with the rapid development of enclosed shopping malls[12] and the emergence of fast food restaurants, shopping patterns changed, and stores in downtown areas closed. Thus began a movement from a monocentric city with centralized economic activity to a polycentric city with economic activity of all kinds dispersed and located within a few miles of the suburban residences (Gordon and Richardson 1997; Ewing 1997, Theobald, 2001). Over the past twenty years, more and more businesses started to locate offices and other facilities closer to the suburban areas where the workers lived. Each suburb assumed a unique character – high income, low income, white collar, blue collar – depending on the employment opportunities that evolved.

This polycentric form of development, in which a central city is surrounded by many different suburbs containing not only residences but office and other business space, and including facilities such as schools, churches, shopping malls and other retail stores now characterizes many suburban areas in the US. A high proportion of suburban residents no longer commute to distant downtown areas in the city core to work, and this limits travel time.

Point-counterpoint perspectives on modern polycentric suburban sprawl can be found in papers by Ewing (1997) and by Gordon and Richardson (1997) and Theobald (2001) proposes a "spawl index" for measuring the extent of urbal sprawl. Gordon and Richardson see modern suburban sprawl as being undesirable and believe that cities should be compact and built on limited acreage with a clear demarcation between what is urban and what is rural. They see low density housing as being wasteful of energy and other resources. The Gordon and Richardson view is consistent with current attempts by leaders in a number of cities to limit urban sprawl (Egan 1997; Williams 1997). For example, Portland, Oregon, with heavy restrictions on the availability of land for development under regulations enacted in the 1970s, represents an experiment in this regard.[13]

Ewing (1997) views contemporary polycentric sprawl with many suburban work and shopping areas that limit commute times far more favorably. He sees polycentric suburban areas as efficient and saving of energy and other resources. Modern telecommunications enable people to work in many outlying areas (suburban office buildings), without needing to commute to a central core. But, there are also impacts (perhaps negative) on social capital formation in situations where workers located at various scattered suburban sites seldom see each other face-to-face. For this reason, places such as New York's financial district or California's Silicon Valley often have a competitive advantage over suburban office parks (Lang, communication).[14]

A suburban development aimed at social capital formation is that of Columbia, Maryland (Christensen 1986). This development is largely consistent with Ewing's vision of polycentric suburbs. A team of behavioral scientists was asked to design a site plan that would provide each resident with a sense of membership within the community. The design created divided the community into a number of one-mile wide neighborhoods each with a population of approximately 3000 residents, and each containing facilities including a school, park, pool, playground and convenience store. Neighborhoods combine to form a village, containing additional facilities such as a middle school, bank, supermarket, pharmacy and additional recreational facilities. Christensen (1986, p.117) quotes one of the planning team consultants as wanting to "break the community down into groups small enough for people to feel an identity and involvement."

## Urban Core

One-way streets designed to facilitate rapid traffic flow are characteristic even of residential areas within our prototype *urban core*. Heavily-traveled one-way streets with comparatively fast-moving traffic impose a physical as well as psychological barrier to the interaction of residents from opposite sides of the street. The buildings are commonly multi-story structures built up to or very near to the sidewalk, with little "front yard" space. This arrangement also means little social interaction among residents living in separate buildings but in close geographic proximity.

Many of the housing units are rentals, some in buildings as small as three stories. These are intermingled with owner-occupied condominiums. There are also high-rise apartment and condominium complexes. Slightly farther out are streets lined with two-story town-houses sharing common walls. Some urban churches remain in operation, but many have exited to the suburbs. In cities without active downtown revitalization programs, residents must increasingly travel outside the urban core for even such basics as groceries. There may be considerable interaction among residents living in the same building, although sometimes even this is limited. Places such as city museums provide some opportunities for social interaction. Residents often have ready access to fine restaurants, a public library, live theater and the like, and these provide opportunities for "non-geographic" social capital formation. School-age children may be bussed to suburban schools located ten or more miles away.

Katz (1994) argues that it is only as recently as the 1980s that significant numbers of people started once again longing for the conveniences traditionally associated with older neighborhoods located in close proximity to central cities. In many smaller cities, this rising interest coincided with the increasing commuting times associated with the new housing developments being built ever further from central cities. These shifts may have had a significant impact on comparative appreciation rates for residential properties since the 1980s in these cities, often leading to gentrification. More specifically, commuting ease and time have long had an important impact on comparative appreciation rates for residential properties in heavily urbanized areas near urban cores in the Northeast and in southern California.[15]

In central-city urban areas, social capital can vary significantly from one residential neighborhood to another, and in many low income, crime-ridden public housing areas, it is clear that productive social capital has completely broken down. Police in urban areas can quickly determine the locales with negative or non-existent social capital, because the bulk of police resources are allocated to those beats. The physical design of housing structures can also influence criminal behaviour, which has recently given rise to the approach of "Crime Prevention Through Environmental Design" (Alberta Community Crime Prevention Association, 1997).

Those of the same race or ethnic group tend naturally to form social capital networks that are stronger than the networks which cross racial and ethnic lines (Warren 1996). With some important exceptions, ethnically diverse urban neighborhoods tend to have lower social capital, since social capital necessarily involves interactions among residents in the neighbor-hoods; Nyden, Maly and Lukehart (1996) discuss this in the context of diverse neighborhoods being perceived as "inherently unstable". These authors (1996, pp. 24-6) list 14 recommendations for "strengthening diverse communities." Christensen (1986) has noted a high degree of racial integration within the neighborhoods and villages of Columbia, Maryland, whereas the degree of socioeconomic integration is not nearly as high. Green (1996, p.7) argues that understanding the role of social capital in stimulating business in rural areas, and determining how policy enhances or reduces this form of capital, will be critical as devolution progresses.

## Research Issues and Questions

Despite all of the interest among researchers in defining and measuring social capital, many important questions remain unanswered. The apparently simple goal of defining social capital and describing its formation remains elusive, apart from the attempts by researchers to define social capital from the perspective of measurable social capital indicators such as measures of volunteerism, membership in civic groups and the like. A contribution of this paper is to show that there are different types of social capital, which are context-specific, and that some forms of social capital are counter-productive to achieving desired social ends. A significant continuing challenge is to distinguish activities which demonstrate "social cohesion" from those which demonstrate "social capital formation"; to separate indicators or proxies of social capital from the social capital itself and; more generally, to separate social capital as *means* from social capital as an *end*. Even so, a research agenda related to social capital and housing is beginning to emerge, as we discuss in these concluding comments.

A first issue worthy of further research is the impact of social capital formation on changes in residential property values. Social capital appears to be a major underlying factor in determining whether a neighborhood is described by its residents as "desirable" or "unsatisfactory" and, *ceteris paribus*, property values tend to rise in what are deemed desirable neighbourhoods, and fall in neighbourhoods considered undesirable. Mortgage lenders clearly have a stake in helping assure that property values are at least stable if not increasing. Local governments rely heavily on revenue from property taxes to fund a variety of services including education, and a decline in property values could lead to a deterioration of these services. Owners of residential real estate are also concerned that their properties appreciate in value.

A second issue is the role of social capital formation in general economic development. Firms prefer to locate plants in areas that workers regard as desirable places in which to live (McLough-lin 1983; Knapp and Graves, 1989). To the extent that local residents in an area are able to form social capital, businesses may increasingly see such an area as a desirable place in which to locate, fostering economic development.

A third issue is the question of whether rising incomes of residents lead to increased social capital formation or, instead, whether increased social capital formation ultimately causes incomes to rise. If there is a cause-and-effect relationship here, what is the direction of causality? Or, which comes first? A related question is how income distribution *within* a community affects social capital formation. For example, if the income distribution is lopsided, with wealthy residents living in closed and gated subdivisions, can social capital still form between those living inside and outside of the gates, and what difference does it make for property values over time?

Last, is an evolving sense of *permanence* to the community necessary for social capital accumulation? If so, how does this sense of permanence manifest itself? Are quantifiable indicators such as residential mobility adequate to measure this factor? Americans, particularly those living in suburban areas, traditionally move to higher-income neighborhoods as they attain additional income. Does this upward mobility that characterizes much of American society inevitably contribute to a decline in social capital formation compared with the relative stability that has at least historically existed in rural communities?

## NOTES

1. We use the terms metro and non-metro and urban and rural interchangeably in this paper (see also footnote 3). According to the U.S. Department of Agriculture, counties with no places containing more than 2,500 people are considered *rural*; counties with places containing urban populations of 2,500 to 20,000 residents are considered to be *non-metropolitan* (and non-rural), while all other counties are considered to be *metropolitan* (Hady and Ross 1990).
2. Webster provides two definitions for the term community–one tied to geography ("a group of people residing in the same locale and under the same government")–but the other is not geography-based ("a group or class sharing common interests"). In a similar manner, Flora (1997, p. 113) refers to "communities of place" and of "interest." Like the definition of social capital, the definition of community that is appropriate depends on the context in which it is used (Bender 1978).
3. Arguably, the function $f$ could itself also depend on the level of social capital in a community. Moreover, $y$, $s$ and $w$ differ depending on the type of community being studied..
4. The terms rural and urban have specific meanings within government agencies such as the U.S. Bureau of the Census. Castle (1995, p. 9), however, notes that the term "rural" can have many meanings, and that even the National Rural Studies Committee struggled to define the term. Castle emphasizes that the various authors in his book are sometimes vague and inconsistent in the use of the term. Lewis, in an essay from Castle's reader, discusses how the rural and the urban are merging in what he calls "the Galactic City".
5. Of course, neighborhoods are made up of families, and so the functioning of families has immediate relevance for the quality of neighborhoods and the value of homes therein.

6. This occurred recently in rural Clinton county, Kentucky, when a chicken processor made plans to locate a processing plant. Local residents objected to the possibility that low-wage workers might be recruited from outside the area, fearing that these workers would possess different values and place new demands on scarce local social services.
7. For more recent discussions in the urban ecology literature of residential segregation by socio-economic characteristics see, *inter alia*, Warren 1996 or Massey and Denton 1993.
8. Blakely and Snyder (1997) estimate that there are almost 19,000 gated communities providing homes to nearly 8.5 million inhabitants.
9. Transportation planners calculate a "connectivity index" to numerically determine how well-connected a street system is. High values are associated with ease of use, whereas low values suggest reduced construction costs. This index is calculated at 1.5 for the prototype rural community illustrated here, but less than 1.0 for the illustrated suburban development, suggesting that the rural community has a greater ease of use—and presumably higher social capital formation potential than for the prototype suburban development.
10. Do those residing in these neighborhoods share values more profound than in their choice of brickwork and brass coach lanterns? Conceivably, part of the satisfaction (utility) for those living in an upscale neighborhood is the ability to socialize with people similarly stationed on the economic and social ladder. Residents of upscale neighborhoods derive satisfaction from knowing they can afford to live in an area that others, by virtue of their lesser economic stations in life, cannot afford, apart from the satisfaction obtained from owning the high-priced home with all its features. While perhaps superficial or shallow, to a certain degree the shared architecture, brickwork and paint schemes imply a shared sense of values.
11. We do not mean to imply that the form of neighborhood necessarily "causes" a lack of interaction. Instead, major lifestyle changes that have contributed to less leisure may be the cause, including both parents working during the week in the case of families, etc.
12. The impact of enclosed shopping malls on social capital formation relative to the type of social capital formation that takes place

in a traditional main street is complex and uncertain, and is sufficiently complicated to warrant a separate paper. Enclosed shopping malls can foster social capital formation as residents meet and socialize at the shopping mall.

13. Twenty-six separate communities formed an elected regional government called METRO to administer the designation of the urban growth boundary (Abbot 1997). The type of social capital giving rise to and reflected in this alliance of 26 communities apparently does not exist in other major cities currently struggling! with limited success with urban sprawl consistent with the Gordon and Richardson model include Salt Lake City, UT; Boulder, CO; Seattle, WA; and Phoenix, AZ. See also the discussion of the undesirable consequences of these kinds of developments in the Florida State Transportation Policy Initiative (1994) describing the Tampa Palms area.

14. The competitive advantage in this case arises from agglomeration economies, which are cost savings that result when firms in the same industry located in close proximity to one another (e.g, Goetz 1997, p.840). We hypothesize that, without social capital, agglomeration economies are reduced significantly.

15. In Lexington, KY, where commuting times from even new construction in outlying subdivisions remain reasonable, unpublished research by the authors shows that values of residential property in zip codes close to the urban core have over the last two decades appreciated in value by a greater percentage than have values of properties located in more recently developed subdivisions located farther from the urban center. Research in progress by the authors suggests that in Lexington, KY, for example, single family residences located within a narrow corridor close to the city center and the University have experienced an average annual appreciation of approximately 5.9 percent per year, whereas similar single family residences located outside the outer loop that circles the city have had an appreciation rate of only 3.2 percent per year, on average. It is not clear to what extent this kind of research can be generalized to other communities; in many metropolitan areas, such as Washington DC or Los Angelos, the commute may be from one suburb (or exurb) to another.

## REFERENCES

Abbot, Carl. 1997. The Portland Region: Where City and Suburbs Talk...And Often Agree. *Housing Policy Debate* 8(1):11-51.

Adler, Jerry, with Douglas Holt, Nina Biddle, and Elizabeth Roberts. 1995. Here Come the Joneses. *Newsweek*. December 5, p. 70.

Alberta Community Crime Prevention Association. 1997. What is Crime Prevention Through Environmental Design? January. At http://www.vtpi.org/tdm/tdm37.htm

Ayres, Janet. 1996. Essential Elements of Strategic Visioning. In: *Community Strategic Visioning Programs*, ed. Norman Walzer. Westport, CT: Praeger.

Beggs, John J., Valerie A. Haines and Jeanne S. Hurlbert. 1996. Revisiting the Rural-Urban Contrast: Personal Networks in Nonmetropolitan and Metropolitan Settings. *Rural Sociology* 61(2):306-325.

Bender, Thomas. 1978. *Community and Social Change in America*. New Brunswick, New Jersey: Rutgers University Press.

Blakely, Edward J., and Mary Gail Snyder. 1997. *Fortress America: Gated Communities in the United States* Washington D.C.; Brookings Institution Press and Lincoln Institute of Land Policy.

Bourdieu, Pierre, 1985. "The Forms of Capital." In: *Handbook of Theory and Reserch for the Sociology of Education*. ed John Richardson, New York: Greenwood. pp. 241-58.

Broom, Leonhard and Philip Selznick. 1963. *Sociology: A Text with Adapted Readings*. New York: Harper & Row, 3rd ed.

Castle, Emery, ed. 1995. *The Changing American Country-side: Rural People and Places*, University Press of Kansas.

Christensen, Carol A. 1986. *The American Garden City and the New Towns Movement*. Ann Arbor: UMI Research Press.

Clinton, Hillary Rodham. 1996. *It Takes a Village: and Other Lessons Children Teach Us*. New York: Simon and Schuster.

Coleman, James S. 1988. Social Capital in the Creation of Human Capital. *American Journal of Sociology* 94 (Supplement 1988):S95-S120.

"Community" 1988. *Webster's II New Riverside University Dictionary*. p. 288 Houghton Mifflin, Boston, MA.

Debertin, David L. 1993. Rural Development Issues for Agricultural Economists to the Year 2000: Discussion. *American Journal of Agricultural Economics* 75 (5):1173-4.

Debertin, David L., Craig L. Infanger and Stephan J. Goetz. 1991. The Viability of Rural Communities: Challenges for the 1990s and Beyond. In *Rural Planning and Develop-ment: Visions of the 21st Century*, (Vol. 1):289-298.

DeFilippis, James. 2001. "The Myth of Social Capital in Economic Development," *Housing Policy Debate*, 12:4. pp. 781-806. At http://www.fanniemaefoundation.org/programs/hpd/v12i4-index.shtml

Egan, Timothy 1997. "Drawing the Hard Line." *Lexington Herald Leader*. Lexington, KY, February 16, p. E2.

Ewing, Reid. 1997. Alternative Views of Sprawl: Counterpoint: Is Los Angeles_Style Sprawl Desirable? *Journal of the American Planning Association* 63(1):107-126.

Fine, Ben. 2001. *Social Capital Versus Social Theory: Political Economy and Social Science at the Turn of the Millenium*. London:Routledge

Flora, Cornelia Butler. 1997. *Community*, In: Gary E. Goreham, editor, *Encyclopedia of Rural Areas*. ABC-CLIO, Santa Barbara, CA. pp. 112-117.

Florida State Transportation and Policy Initiative. 1994. Evaluation of Community Types. In: *Transportation, Land Use and Sustainability*. Florida Center for Community Design and Research at http://www.fccdr.usf.edu/projects/tlushtml/tlus110.htm

Freeman, Scott, Jeffrey Grogger and Jon Sonstelie. 1996. The Spatial Concentration of Crime. *Journal of Urban Economics* 40: 216-231.

Goetz, Stephan J. 1993. Human Capital and Rural Labor Issues. *American Journal of Agricultural Economics* 75: 1164-1168.

Goetz, Stephan J. 1997. State- and County-Level Determinants of Food Manufacturing Establishment Growth: 1987-93. *American Journal of Agricultural Economics* 79:838-850.

Goetz, Stephan J. and David L. Debertin. 1996. "Rural Population Decline in the 1980s: Impacts on Farm Structure and Federal Farm Programs." *Am. J. Agr. Econ.* 78(3):517-29.

Gordon, Peter, and Harry W. Richardson. 1997. Alternative Views of Sprawl: Point: Are Compact Cities a Desirable Planning Goal? *Journal of the American Planning Association* 63(1):95-106. Available in the Smart Growth Library at http://www.smartgrowth.org/library/ apa_ pointcounterpoint/apa_sprawl.html

Green, Gary P. 1996. Social Capital and Entrepreneurship: Bridging the Family and Community. Paper presented at the Cornell University Conference on Entrepreneurial Families—Building Bridges, New York, March 17-19. At http://www.fambiz.com/Orgs/Cornell/articles/real/green.cfm

Grigsby, W.G. 1963. *Housing Markets and Public Policy*. Philadelphia: University of Pennsylvania Press.

Grigsby, W.G., M. Baratz, G. Galster and D. MacLennen. 1987. *The Dynamics of Neigh-borhood Change and Decline*, Oxford: Pergamon Press.

Hady, Thomas F. and Peggy J. Ross. 1990. An Update: The Diverse Social and Economic Structure of Nonmetropolitan America. Staff Report No. AGES 9036, United States Department of Agriculture, Washington, DC, May.

Helliwell, John and Robert Putnam. 1995. Economic Growth and Social Capital in Italy. *Eastern Economic Journal* 21:295-307.

Hirschfield A. and K.J. Bowers. 1997. The Effect of Social Cohesion on Levels of Recorded Crime in Disadvantaged Areas. *Urban Studies* 34(8):1275-1295.

Hornburg, Steven P. 1998. *Personal Correspondence*, dated February 27.

Hughes, William T. and Geoffrey K. Turnbull. 1996. Uncertain Neighborhood Effects and Restrictive Covenants. *Journal of Urban Economics* 39:160-172.

Katz, Larry. 1992. Recent Development in Labor Economics. Amer. Econ. Association lecture on Recent Developments in Economics; New Orleans, LA, annual meeting.

Katz, Peter. 1994. *The New Urbanism: Towards an Architecture of Community*, New York: McGraw-Hill.

Kenworthy, Lane. 1997. Civic Engagement, Social Capital, and Economic Cooperation. *American Behavioral Scientist* 40(5):645-656.

Knack, Stephen and Philip Keefer. 1997. Does Social Capital Have an Economic Payoff? A Cross-Country Investigation. *Quarterly Journal of Economics* 112(4):1251-1288.

Knapp, T.A. and P.E. Graves. 1989. On the Role of Amenities in Models of Migration and Regional Development. *Journal of Regional Science* 29 (Feb.):71-87.

Laird, Donald M. 1996. Geographical Changes in *Money* Magazine's "Best Places to Live" Rankings. *Economic Development Review* Spring:24-28.

Lang, Robert E. 1998. Personal Correspondence, dated March 24.

Lang, Robert E., and Karen A, Danielson. 1997. "Gated Communities in America: Walling Out the World 8(4) : 867-877.

Lewis, Pierce. 1995. The Urban Invasion of Rural America: The Emergence of the Galactic City. In: *The Changing American Country-side: Rural People and Places,* ed. Emery Castle, ed. 39-62. University Press of Kansas.

Massey, Douglas S. and Nancy A. Denton. 1993. *American Apartheid: Segregation and the Making of the Underclass*, Cambridge: Harvard University Press.

McLoughlin, P.J. 1983. Community Considerations as Location Attraction Variables for Manufacturing Industry. *Urban Studies* 20:359-63.

Megbolugbe, Isaac F., Marja C. Hoek-Smit and Peter D. Linneman. 1996. Understanding Neighborhood Dynamics: A Review of the Contributions of William G. Grigsby. *Urban Studies* 33 (10):1779-1795.

Mesch, Gustavo S. 1996. The Effect of Environmental Concerns and Governmental Incen-tives on Organized Action in Local Areas. *Urban Affairs Review* 31(3):346-366.

Moore, Carl M. 1991. Community is Where Community Happens. *National Civic Review* 80(4):352-357.

Newton, Kenneth. 1997. Social Capital and Democracy. *American Behavioral Scientist* 40(5):575-586.

Nyden, Phillip, Michael Maly and John Lukehart. 1996. The Emergence of Stable Racially and Ethnically Diverse Urban Communities: A Case Study of Neighborhoods in Nine U.S. Cities. Paper presented at Fannie Mae Foundation Annual Housing Conference, May 10.

O'Looney, John. 1993. Framing a Social Market for Community Responsibility: Governing in an Age of NIMBYs and LULUs. *National Civic Review* 82(1):44-62.

Ostrom, Elinor. 1994. Constituting Social Capital and Collective Action. *Journal of Theoretical Politics* 6(4):527-562.

Petee, Thomas A. and Gregory S. Kowalski. 1993. Modeling Rural Violent Crime Rates: A Test of Social Disorganization Theory. *Sociological Focus* 26(1):87-89.

Portney, Kent E. and Jeffrey M. Berry. 1997. Mobilizing Minority Communities. *American Behavioral Scientist* 40(5):632-644.

President's Council on Sustainable Development. 1995. Sustainable America: A New Consensus. At http://clinton2.nara.gov/PCSD/Publications/TF_Reports/amer_top.html

Putnam, Robert D. 1993*a*. The Prosperous Community: Social Capital and Public Life. *The American Prospect* 13 (Spring), at http://www.epn.org/prospect/13/13putn.html

Putnam, Robert D. 1993*b*. *Making Democracy Work: Civic Traditions in Modern Italy*. Princeton, NJ: Princeton University Press.

Putnam, Robert D. 1995. Bowling Alone: America's Declining Social Capital. *Journal of Democra-cy* 6: 65-78. Also at http://lwv.org/~lwvus/putnam.html

Putnam, Robert D. 2000. *Bowling Alone: The Collapse and Revival of American Community*. New York: Simon and Schuster.

Ray, Brian K., Greg Halseth and Benjamin Johnson. 1997. The Changing 'Face' of the Suburbs: Issues on Ethnicity and Residential Change in Suburban Vancouver. *International Journal of Urban and Regional Research* 21(1):75-99.

Rubio, Mauricio. 1997. Perverse Social Capital!Some Evidence from Columbia. *Journal of Economic Issues* 31(3):805-816.

Salamon, Sonya. 1995. The Rural People of the Midwest. In *The Changing American Countryside: Rural People and Places,* ed Emery Castle, University Press of Kansas, 352-65.

Schmid A. Allan and Lindon J. Robison. 1995. Applications of Social Capital Theory. *Journal of Agricultural and Applied Economics* 27(1):59-66

Sirianni, Carmen and Lewis Friedland. 1997. Social Capital. Civic Practices Network, at http://www.cpn.org/tools/dictionary/capital.html

Teachman, Jay D., Kathleen Paasch and Karen Carver. 1997. Social Capital and the Generation of Human Capital. *Social Forces* 75(4):1343-59.

Theobald, David M. "Quantifying urban and rural sprawl using the sprawl index" Colorado State University, paper presented at the annual conference of the Association of American Geographers, New York, New York, March 2nd, 2001. Available at http://ndis.nrel.colostate.edu/davet/pubs%5Caag2001_v2.htm

Warren, Mark R. 1996. Creating a Multi-Racial Democratic Community: A Case Study of the Texas Industrial Areas

Foundation. Adapted from Ch. 4 of *Social Capital and Communi-ty Empowerment: Religion and Political Organization in the Texas Industrial Areas Foundation.* Ph.D. dissertation., Harvard Univ. 10. At <http://www.cpn.org/sections/topics/community/stories-studies/texas_iaf1.html>

Weiss, Michael J. 1989. *The Clustering of America.* New York: Harper and Row.

Williams, Larry. 1997. Trading Farmland for Front Yards? *Lexington Herald Leader*, March 21, pp. A13.

Wills, Garry. 2000. Putnam's America. *American Prospect* 11:16 Available at http://www.prospect.org/print/V11/16/wills_g.html

Wilson, Patricia A. 1997. Building Social Capital: A Learning Agenda for the Twenty-first Century. *Urban Studies* 34(5-6):745-760.

# 9

# The Role of Social Capital within Regeneration: Can Building Social Capital Benefit Regeneration Contexts?

*Louise Warwirk-Booth*

## Abstract

This review of the literature examines social capital in relation to regeneration, theoretically drawing out the potential links between the concept of social capital and its relationship to regeneration.

Examining the work of Bourdieu, Coleman, Putnman and Fukuyama, different conceptualisations of social capital are discussed in order to understand how it might be operationalised within regeneration contexts.

The review then discusses the benefits of linking social capital to regeneration and finally moves onto some of the issues that may arise with the use of the concept before concluding that despite many theoretical links and many potential contributions, social capital building within regeneration requires a carefully considered approach.

**Key words:** *social capital / regeneration / Bourdieu / Coleman / Putnam / Fukuyama / grassroots development*

## Introduction

In recent policy initiatives community involvement, community participation and increasingly social capital have been the subject of discussion and debate with reference to social regeneration. This article explores the role of social capital within regeneration, discussing the potential links between the concept and development through a critical review of the existing literature in order to map the relationship between social capital and regeneration.

## Context

The emergence of the concept of social capital within the sociological literature and within policy debates is relatively recent. However, the academic and policy-making communities have been energised by the concept because of its implications for development policy and its potential political contribution. Social capital demonstrates how the presence of dense networks within society and accompanying norms of generalised trust and reciprocity allow people to overcome collective action problems more efficiently (Hooge and Stolle 2003). As a result, the concept has been embraced as a solution for a plethora of social problems (Boix and Posner 1998). For example, in the context of the 'Third World' social capital has directly entered into the policy discourse of the World Bank, with the Bank describing it as the missing link in development (Harriss & de Renzio 1997). Furthermore, at the European level there has been an initiative to develop 'local social capital' because of the recognition of its role within regional development.

Within the UK social capital as a concept was brought into the policy arena by the Commission for Social Justice. Drawing heavily on the work of Putnam, the Commission defined the concept in terms of community strength, civic wealth and linked it to the geography of the neighbourhood (Commission for Social Justice 1998). These ideas persisted and were included in the work of the Social Exclusion Unit in 1998, which detailed the government's strategy for neighbourhood renewal. Social capital is highly visible within the UK policy arena and arguably has much to offer social and urban regeneration. Indeed, if positive social outcomes emerge from social capital as a resource and if social capital production can be enhanced then social policy could harness it in order to effect change (Gamarnikow and Green 1999). However, some commentators argue that despite this widespread attention, there is still arguably a limited theoretical understanding of the concept, which remains in its infancy (Boix and Posner 1998). Firstly, let's turn to the question of what is meant by the concept.

## Definitions

So, what is social capital? Although social capital is a relatively recent concept, it is well developed in theoretical terms with its roots being visible in the work of Tonnies', Durkheim and Weber (Woolcock 1998). Social capital for Durkheimians is a form of normative contract making. The concept makes it possible for individuals to commit to action, shape goals and attain them. Comparatively, understanding social capital in Weberian terms means construing the concept as a combination of ties and norms that bind together individuals within and across institutions. Furthermore, much of the social capital literature harbours an implicit concern for the loss of Tonnies 'gemeinschaft' with the solution to modern life being the re-establishment of local civic organisations (Woolcock 1998). These different conceptualisations suggest that social capital may exist in a number of forms. All definitions of the concept suggest that it has positive consequences for members of groups, achieved through shared norms, networks and trust. So how are different conceptualisations relevant to regeneration?

## Pierre Bourdieu

Bourdieu during the 1960s and 1970s produced a series of studies seeking to establish culture as dynamic and creative but also as a structured phenomenon. Bourdieu's understanding developed the concept as a critical tool to help explain poverty in class terms (Everingham 2003) and this places the focus at the level of the individual. Bourdieu described social capital as a potential resource linked to networks and relationships of mutual acquaintance and recognition, giving members collectively owned capital (Schuller *et al.* 2000). More recently, Bourdieu (1999) identified the superimposition of social and physical space and the associated disadvantages that are bestowed upon less powerful social groups by their residence in poor areas in which collective decline has resulted from industrial decline. The concept here is used to explain the privileged access that some individuals have to both powerful groups and their resources.

Bourdieu's concept is rooted in Marxist tradition, emphasising the structural constraints on individuals and the unequal access that people have to resources based on class, gender and race (Everingham 2003). For Bourdieu, access to resources and issues of power within society are the key to social capital (Harper 2001). Social capital effectively viewed as connections and social obligations in Bourdieu's (1986) understanding can be converted under certain conditions into economic capital. This notion has greatly appealed to policy makers as it appears a quick-fix solution but this fails to focus upon the complexity of the 'conditions' under which such a process actually happens (Leonard 2004). There is however growing consensus that area effects do exist. Reviews of American research (see Brooks-Gun *et al.* 1993, Ellen and Turner 1997 and Jencks and Mayer 1990) all conclude that there are causal connections between poor neighbourhoods and other social problems. Galster and Zobel (1998) argue that these result from more than the consequences of household and individual characteristics and if this is the case then social capital is crucial within regeneration. As Atkinson and Kintrea (2004) argue, if individuals in deprived areas are held back because of where they live rather than by their individual characteristics, this is important in formulating urban policy.

What is important from this perspective is the way in which networks operate to enhance or constrain success. However Bourdieu's focus at the level of the individual is less useful because regeneration tends to focus on a macro level such as structural inequalities. Now let's turn to another view.

## Coleman

Coleman's (1998) approach is to define social capital by its function, viewing it as a resource to be drawn upon. He argues that social relations constitute useful capital resources for actors through processes such as obligations, expectations, trust, information channels and setting norms. According to Coleman

(1990) social capital takes three forms. Firstly, the obligations and expectations which depend upon the trustworthiness of the social environment. Secondly, the capacity of information to flow through the social structure in order to provide a basis for action. Finally, it is the presence of norms accompanied by effective sanctions. Social capital differs from other forms of capital in that it does not necessarily bring benefits just to the individual rather it brings benefits to all of those who are part of the social structure. Individuals are drawn into social structures through obligations and effective sanctioning ensures that co-operation occurs. Furthermore, social capital in both family and community plays a role in creating human capital such as skills and abilities within the next generation (Leonard 2004).

Despite this recognition of the wider benefits of the concept, Coleman's theorising primarily focuses upon individuals and the family. Coleman also argues that social capital creation is a largely unintentional process (Schuller *et al.* 2000) and that although it is productive, it can be depleted if it is not renewed. Hence, if social capital production is unintentional, how can it be encouraged via social policy or harnessed for regeneration purposes? Furthermore, Coleman's concept, if applied in practice would lead researchers to focus upon social capital as a positive good within society that is necessary for both social integration and social control (Everingham 2003). Critics say that this approach overstates the importance of closure and dense ties within the social structure and treats social capital in an unproblematic manner (Schuller *et al.* 2000). However, Coleman (1998) does discuss the detrimental effect that social capital can have by discussing different forms, which can not necessarily be aggregated. The enabling of some forms of social capital may simultaneously inhibit the formation of other types.

Coleman's view of social capital again can contribute to understanding regeneration particularly as his gaze turns to the community benefits of the concept. If social capital benefits the community as a whole and can be developed through effective regeneration, it could serve as a useful tool in development work. This is especially true if social capital creates human capital because the enhancement of different forms of skill within deprived areas is the goal of much regeneration work. However, the uncritical stance taken by Coleman suggests caution. Even if human capital can be increased, this does not necessarily ensure better economic prospects because structural factors can serve to impede economic development.

## Putnam

Putnam's work on social capital popularised the concept enabling it to find its way into mainstream political discourse (Schuller *et al.* 2000). According to Putnam (2000) social capital refers to the connections among individuals, social networks and other forms of reciprocity and trust which arise from them. Networks, norms and trust dominate his definition of the concept whilst activity is situated at the heart of civic life and therefore, is also a crucial aspect of his conceptualisation (Schuller *et al.* 2000). Putnam (1993) suggests that the more people work together, the more that social capital is produced. Putnam (1993) argues that social capital; cooperation, trust, reciprocity and collective well-being define the civic community, which has deep historical roots. Thus, social capital according to Putnam (1993, 2000) is the key to economic growth and the key to making democracy work within the context of civic society. Putnam suggests that economic growth flows from social progress and that stocks of capital can be accumulated at the institutional level and passed on. Putnam's perspective concurs with the political science view of the concept as crucial in civic engagement and trust at the level of the community and how these foster the development of civic trust in political institutions (Leonard 2004).

Putnam transforms social capital from a 'simple' social process, which benefits individuals and groups, to an entity that can be consolidated for society's benefit as a whole because of the links between reciprocation, democracy and economic growth (Everingham 2003). However, this definition can slide into a circular argument where the mechanisms of social capital become confused with the outcomes so that researchers employing this framework end up finding whatever it is that they set out to find (Stone 2001). Furthermore, Putnam's work imbues community with highly positive connotations portraying an image of helpful, friendly interactions between individuals based upon personal knowledge and face to face contact. This ignores the downside of community life (Leonard 2004).

Despite these criticisms, Putnam's view of social capital is relevant to economic regeneration with his formulation evident in the development literature, driving economic growth and potentially alleviating poverty (Everingham 2003). However, this will only occur when engagement takes place in secondary associations, according to Putnam (1993). Although Putnam's (1993) interpretation is good politically, it may not offer insights at the micro-level as to why difficulties with engagement in area based regeneration take place.

## Fukuyama

Fukuyama primarily presents social capital as trust by defining the concept as 'a set of informal values or norms shared amongst members of a group that permits co-operation between them' (1999:16). Most social capital definitions pay attention to trust and give equal weight to trust and networks but some prioritise one over the other (Berman and Phillips 2003). For Fukuyama (1999) the most important value is trust. It is trust that leads to co-operation and therefore makes both groups and networks operate smoothly. Central to this conceptualisation is the radius of trust, where it is argued that the further trust expands outside of the family then the more likely it is to be based upon moral resources and ethical behaviours (Fukuyama 2001). Where groups have a narrow radius of trust, their in-group solidarity reduces their ability to co-operate with outsiders. It is arguably difficult for people to trust those outside of narrow circles especially in the absence of weak ties. This argument about in-group trust reflects parallels to Putnam's emphasis on close, tight-knit networks not always being beneficial.

Fukuyama (1999) contributes to social policy discussions in arguing that states do not have many obvious levers by which to create social capital. He asserts that states can have a negative impact upon social capital development if they undertake certain activities that are better left to the private sector or to civil society because of the creation of state dependency. Therefore grass-roots regeneration has more to contribute to social capital development than top-down governmental approaches according to this interpretation. However, Fukuyama's (1999) approach has again been criticised because of his monoculturalist standpoint in which he asserts that societies need to share the same language, norms and moral values in order to avoid disintegration. Other commentators for example Kymlicka (1995) attempt to offer a more multiculturalist perspective. Despite criticism, Fukuyama's interpretation can shed light on regeneration practice. Communities in which high levels of poverty exist can be insular and have increased levels of in-group trust, which serves to exclude others. Therefore development work needs to create a wider radius of trust and arguably this is more achievable through grass-roots approaches.

## Discussing Definitions

These definitions of social capital all highlight elements such as trust and associational linkages, although they give different weight to their importance. The different definitions also emerge from a variety of sociological traditions. Bourdieu draws upon Marxism, Coleman is concerned with function, Putnam politically locates his interpretation and Fukuyama's discussion is inherently conservative. Despite these different traditions all of these interpretations suggest that the concept is useful within regeneration. Drawing through functionalism and political science with a community and economic focus allows the relevance of social capital within regeneration to be broadly explained. For example, both Coleman and Putnam discuss the macro benefits of social capital. Coleman's functionalist underpinnings lead to a focus on the community level benefits of the concept. If social capital can provide community level benefits then it is useful as a regeneration driver, effectively meeting the social aspect of development work. Putnam's focus is more economic, with the benefits of the concept described in development terms as driving growth and alleviating poverty. This is also pertinent to regeneration as development work often has an economic focus.

## Social Capital within Regeneration

Social capital is indeed recognised within the wider regeneration literature for its potential contribution. MacGillivray and Walker (2000) discuss the relationship of social capital to sustainability because networks can be used to pull individuals and communities out of poverty. Hibbitt *et al.* (2001) conclude that social capital is important in conceptualising area based regeneration initiatives through judging linkages and networks which serve to create effective regeneration.

Social capital is arguably relevant to the voluntary sector in more ways than one. Voluntary organisations contribute to social capital because they are primarily social networks where collective action takes place but they also benefit from existing social capital and use social capital as a resource. For example, social capital as a concept can help voluntary organisations improve their understanding of their area in terms of networks and structures. Social capital also can be used in evaluating activities through the lens of the concept to analyse processes and practices rather than just outcomes (Jochum 2003). Indeed, Cattell (2004) argues that the conditions required to promote social capital such as the built environment and opportunities for participation can include those which regeneration agencies address. So how might policy debates be translated into practical reality at a grassroots level?

## Grassroots Development and the Formation of Social Capital

In a broad sense social capital can be linked to regeneration settings in a number of ways. For example, involvement and empowerment are central to much regeneration practice; similarly associating together and engaging in community affairs are crucial to social capital development for Coleman (1998). Putnam (2000) also discusses participation and reciprocity, therefore the concept may be useful in explaining collective action in terms of mutual involvement and the creation of alliances to achieve group and community goals. Collective action is considered difficult to theorise and explain because cooperation sometimes occurs in contexts where, according to theory, social actors should not engage in it. For example, groups have acted collectively within deprived areas where high levels of crime exist. The concept of social capital has been used to explain this (Boix and Posner 1998), in terms of the existence of both trust and networks. This remains relevant to social regeneration initiatives, which may require collective action to achieve results.

Secondly, social capital arguably greases the wheels of communities in that it enables them to run more smoothly. Putnam (1993) argues that where people are trusting and trustworthy and regularly interact with those around them, social transactions are less costly, greatly facilitating social relationships. Putnam (2000) marshals evidence to demonstrate that in high social capital areas public spaces are cleaner, people are friendlier and the streets are safer. If this is the case then social capital is a great resource to harness and use in development. Indeed, local volunteers can play a role in greasing the wheel because they put time into service delivery, empower themselves and others, free up paid workers to carry out other tasks and increase their own employability. Indeed, some of the UK policy responses within regeneration have focused upon volunteering and community self-help. New Labour's Coalfield Policy promoted a greater degree of community involvement in and ownership of regeneration activities (Waddington 2003). Thus aiming to build social capital through the strengthening of local community networks (Harper 2001) is directed by social policy and related interventions. However, an examination of the literature points to the complex and subtle ways in which inequality manifests itself in community relationships, which on the surface seem to be based upon trust and reciprocity (Leonard 2004). For example, inclusion and exclusion occurs based upon gender, ethnicity, age and religious affiliation. Therefore a cautious approach is required. Policy makers and regeneration drivers need to know what conditions lead to beneficial outcomes in order to orient policy.

Thirdly, social capital has also been argued to have an economic benefit. This is because the mechanisms by which civic values influence socio-economic performance reduce transaction costs in the market, increase trust, minimise burdens of enforcing and policing agreements and hold down the diseconomies of fraud and theft (Putnam 1993 and Coleman 1990). In both Putnam's (1993) and Coleman's (1990) view, social capital is given equal weight to economic and cultural capital and both suggest that social capital paves the way for the acquisition of other forms of capital (Leonard 2004). However, a number of studies have demonstrated how areas rich in social capital habitually under-perform economically compared to other regions (Matthews 1983, Richling 1985). For example, in the case of 'niche' economies, kinship and friendship ties operate to enable individuals to make insufficient incomes stretch to meet their daily needs. However, these networks can also provide access to highly exploitative employment (Leonard 2004). Hence, economic benefits are not always as clear cut as some theorists suggest.

Finally, at the level of the individual social capital has been argued to improve quality of life via psychological and biological processes. Individuals rich in social capital with high levels of social, economic cultural and collective resources cope better with traumas and fight illness more effectively according to Putnam (2000). Social capital can act as a buffer against economic disadvantage by reducing the effects of a lack of economic resources (Campbell 1999). Social capital is said to improve individual awareness of the ways in which the human fate is linked, allowing people to become more tolerant, less cynical and more empathetic to the needs of others. Without these opportunities, facilitated by networking and social relationships, people are more likely to be swayed by their worst impulses (Putnam 2000).

## Overview

Social capital seems to offer a number of benefits to regeneration initiatives. Schuller *et al.* (2000) argue that one of the key merits of social capital as a concept is that it shifts the focus of analysis from the behaviour of individual agents to the patterns of relations between agents, social units and institutions. Thus, social capital arguably acts as a link between the micro, meso and macro-levels of analysis. This allows the concept to make contributions in wider social policy terms. Furthermore, social capital as a concept is multi-disciplinary as well as interdisciplinary; consequently debates about the concept can be found throughout the social sciences. This offers ground for sustained dialogue and future theoretical development within the social science arena (Schuller *et al.* 2000), creating research ties and opening up networking opportunities between disciplines. Finally, the concept has been argued to reinsert issues of value into the heart of social science discourse. Terms such as trust, sharing and community directly generate questions about the assumptions concerning human behaviour on which analysis and policy are based (Schuller *et al.* 2000) offering new insights into regeneration and possibly other policy initiatives. The need to generate norms of reciprocity, to build stocks of social capital, to produce social cohesion and counter the dislocating effects of globalisation is now recognised.

## Problems

Despite these potential links and the surge of policy interest in social capital, there are a number of criticisms of the concept cited within the literature, which will impact and influence the concept in all settings including regeneration. Several theorists argue that social capital as a concept is nothing new and that it is simply being exported wholesale from America to the UK, which ignores the cultural context of its conceptualisation within research studies (Harper 2001). Davies (2002) suggests that the concept is gender blind and ethnocentric whilst other theorists recognise that it is narrow in its focus (Walker and Wigfield 2003). These are just some of the broader criticisms of social capital, other general criticisms focus upon definition (see Schuller *et al.* 2000, Everingham 2003 and Portes and Landolt 1996), precision (see Flora 1998, Everingham 2003 and Wall *et al.* 1998), measurement (see Portes and Landolt 1996, Green *et al.* 2000 and Walker and Wigfield 2003), theoretical underpinnings and epistemology (see Portes and Landolt 1996, Hooge and Stolle 2003 and Garmarnikow and Green 1999).

## Specific Regeneration Problems

This range of general issues associated with social capital may well have an influence within social regeneration settings.

Furthermore, there may be issues particularly pertinent to regeneration, which may also create problems in forming and fostering social capital.

Firstly, trust within any neighbourhood is not guaranteed. The impact of historical divisions within areas, contemporary housing policies, intense deprivation and the sudden presence of streams of money can all act to undermine levels of trust between individuals and groups within neighbourhoods (Hibbitt *et al.* 2001).

Secondly, building social capital may not be suitable for all regeneration contexts or purposes. The literature recognises that tailored and integrated responses are necessary in addressing neighbourhood problems. Social capital is highly context dependent (Jochum 2003), because different neighbourhoods have different combinations of factors that affect how they work. Consequently, different factors in different places create success (Groves *et al.* 2003). It is at the level of the neighbourhood that external factors such as demography and geography are the most determining. Consideration must also be paid to the socio economic situation and the institutional environment within any area (Jochum 2003).

Thirdly, Leonard (2004) argues that the endurance of social capital depends on the ability of certain individuals to mobilise their own cultural and economic capital and that of others. Some individuals are better placed within a community to lead others forward in developing social capital. Despite this suggestion, the literature on social capital has largely ignored the importance of leadership. Purdue (2001) suggests that community leaders play a crucial role in accumulating internal social capital through their work at the grassroots level and are also at the forefront of developing external social capital through partnerships with outside elite groups. Within regeneration these key leaders are often dedicated professionals whose capacity building expertise allows local people to undertake and complete such tasks. However, not all areas have good, experienced and capable workers. Many regeneration professionals are both temporary and migratory, therefore they do not engender the type of commitment to the neighbourhood that residents often do. As such, they are not a resource that forms the fabric of neighbourhood social capital in many instances but rather a fleeting resource whose primary commitment is to their employers and themselves (Fraser and Lepofsky 2004).

Fourthly, voluntary organisations especially volunteer led ones often rely heavily upon a small number of people. This reliance should raise concerns not only about dependency and sustainability, but also about inclusiveness and diversity. If voluntary organisations are a source of social capital and contribute to building social capital, the question must be asked about whose social capital they are developing? If such groups wish to produce positive outcomes then they need to encourage diversity and inclusiveness (Jochum 2003). Too much bonding social capital between volunteers and high levels of in-group trust will serve to negate any positive benefits emerging from voluntary associations as exclusivity is likely to be the outcome. Communities and voluntary organisations require all three types of social capital to ensure sustainable development (Stone and Hughes 2001).

Finally, social capital is only valuable to the extent that community members recognise and sustain its value. It may be the case that the use of social capital terminology brings out a negative reaction from people working within voluntary and community organisations (Begum 2003), leaving the concept outside the voluntary arena at least for the imminent future.

## Conclusion

Social capital as a conceptual entity then is highly debated within the sociological literature, wider academic fields and within policy debates. Increasingly, within the policy arena, initiatives are employed which attempt to develop local social capital. Both Coleman (1998) and Putnam (2000) discuss the macro benefits of social capital and both interpretations suggest that on a broad level social capital is useful within regeneration settings. Indeed, some commentators emphasise the links between the concept and regeneration (MacGillivray and Walker 2000).

This review of the literature highlights several theoretical links between regeneration and the development of social capital with the concept appearing to have much to offer social regeneration initiatives. For example, Hibbert *et al.* (2001) carried out a case study research project in Merseyside looking at the Objective 1 Programme in relation to social capital formation within regeneration. They concluded that firstly, social capital is important in conceptualising area based regeneration initiatives and secondly that the regeneration programme they examined was strengthening social capital within neighbourhoods and building relations of trust between members of local communities. Thus, social capital can impact positively within some regeneration settings.

However, social capital as a concept is open to debate and criticisms, not unlike many other social science conceptual tools, much other social policy terminology and indeed empirical evidence within these fields. Thus, social capital creation in the arena of social policy and urban regeneration needs both a considered and critical approach.

## REFERENCES

Atkinson, R and Kintrea K (2004) 'Opportunities and Despair, It's All There; Practitioner Experiences and Explanations of Area Effects and Life Chances' *Sociology* 38(3): 437-455.

Begum, H (2003) *Social Capital in Action: Adding Up Local Connections and Networks. A Pilot Study in London* Centre for Civil Society and NCVO, London.

Berman, Y & Phillips D (2003) *Indicators for Social Cohesion* Paper submitted to the European Network on Indicators of Social Quality of the European Foundation on Social Quality, Amsterdam.

Bourdieu, P (1986) 'The Forms of Capital' in Richardson J (Eds) *Handbook of Theory and Research for the Sociology of Education* Greenwood Press, New York.

Bourdieu, P (1999) *The Weight of the World: Social Suffering in Contemporary Society* Cambridge University Press: Cambridge.

Brocks-Gun, JG, Duncan G, Klebanov, P & Sealane, N (1993) 'Do Neighbourhoods Influence Child and Adolescent Development?' *American Journal of Sociology* 99, 353-93

Boix, C & Posner, DN (1998) 'Social Capital: Explaining its Origins and Effects on Government Performance' *The British Journal of Political Science* 29(3): 686-693.

Campbell, C (1999) *Social Capital and Health* Health Education Authority, London.

Catley, A (2000) 'The Use of Participatory Appraisal by Veterinarians in Africa' *Review Science Technology* 19(3): 702-14.

Cattell, V (2004) 'Having a Laugh and Mucking in Together: Using Social Capital to Explore Dynamics Between Structure and Agency in the Context of Declining and Regenerated neighbourhoods' *Sociology* 38(5): 945-963.

Coleman, JS (1990) *Foundations of Social Theory* The Belknap Press of Harvard University Press, Harvard.

Coleman, JS (1998) 'Social Capital in the Creation of Human Capital' *American Journal of Sociology* Supplement 94: 95-120.

Commission for Social Justice (1998) *Social Justice: Strategies for National Renewal* Vintage, London.

Davies R (2002) 'Monitoring and Evaluation NGO Achievements' in Desai V & Potter RB (Eds) *The Companion to Development Studies* Arnold, Great Britain.

Ellen I & Turner M (1997) 'Does Neighbourhood Matter? Assessing Review Evidence' *Housing Policy Debate* 8(1): 833-66.

Everingham C (2003) *Social Justice and the Politics of the Community* Ashgate, England.

Flora JL (1998) 'Social Capital and Communities of Place' in *Rural Sociology* 63(4), 481-506

Fraser J & Lepofsky J (2004) 'The Uses of Knowledge in Neighbourhood Revitalization' *Community Development Journal* 39 (1): 4-12.

Fukuyama F (1999) *Social Capital and Civil Society* Conference Paper prepared for the IMF Conference on Second Generation Reforms

Fukuyama F (1999) *The Great Disruption. Human Nature and the Reconstruction of the Social Order* Profile Books, London

Fukuyama F (2001) 'Social Capital, Civil Society and Development' *Third World Quarterly* 22: 7-20

Galster G & Zorbel A (1998) 'Will Dispersed Housing Programmes Reduce Social problems in the US? *Housing Studies* 13 (5): 605-22

Gamamikow E & Green A (1999) 'Developing Social Capital: Dilemmas, Possibilities and Limitations in Education' In: Hayton A (Eds), *Tackling Disaffection and Social Exclusion*, Kogan Ltd, London

Granovetter M (1973) 'The Strength of Weak Ties' *American Journal of Sociology* 78: 1360-80

Green, G, Grimsley M, Syolcas A, Prescott M, Jowitt T and Linacre R (2000) *Social Capital, Health and Economy in South Yorkshire Coalfield Communities* Centre for Regional Economic and Social Research, Sheffield Hallam University.

Groves R, Middleton A, Murie A & Broughton K (2003) *Neighbourhoods That Work. A Study of the Bournville Estate, Birmingham* The Policy Press, Bristol.

Hall PA (1999) 'Social Capital in Britain', *British Journal of Politics* 29: 417-461.

Harper R (2001) *The Measurement of Social Capital in the United Kingdom* ONS, London.

Harriss J & De Renzio P (1997) 'Missing link or Analytically Missing: The Concept of Social Capital' *Journal of International Development* 9(7): 919-937.

Hibbitt K, Jones P & Meegan R (2001) 'Tackling Social Exclusion: The Role of Social Capital in Urban Regeneration on Merseyside – From Mistrust to Trust' *European Planning Studies* 9(2): 141-161.

Hills M & Mullet J (2000) *Community-Based Research: Creating Evidenced-Based Practice for Health and Social Change* Paper presented at the Qualitative Evidence-Based Practice Conference, Coventry University, May 15-17, 2000.

Hooge M & Stolle D (2003) *Generating Social Capital. Civil Society and Institutions in Comparative Perspective*, Palgrave Macmillan, Hampshire.

Jencks C & Mayer s (1990)'The Social Consequences of Growing Up in a Poor Neighbourhood' In: Lynn L & McGeary M (Eds) *Inner City Poverty in the United States* National Academic Press, Washington DC.

Jochum V (2003) *Social Capital: Beyond the Theory,* NCVO Publications, London.

Kymlicka W (1995) *The Rights of Minority Cultures*, Open University press, Oxford.

Leonard M (2004) 'Bonding and Bridging Social Capital: Reflections from Belfast' *Sociology* 38 (5): 927-944.

MacGillivray A & Walker P (2000) 'Local Social Capital: Making it Work on the Ground' In: Baron S, Field J & Schuller T (Eds), *Social Capital: Critical Perspectives* Oxford University Press.

Matthews R (1983) *The Creation of Regional Dependency*, University of Toronto Press, Toronto.

Narayan, D (1999) *Bonds and Bridges: Social Capital and Poverty,* World Bank, Washington DC.

Portes A (1998) 'Social Capital: Its Origins and Applications in Contemporary Sociology' *Annual Review of Sociology* 24: 1-24.

Portes A & Landolt P (1996) 'Unsolved Mysteries; The Tocqueville Files II. The Downside of Social Capital' *The American Prospect* 7, 26.

Prusak L & Cohen D (2001) 'How to Invest in Social Capital' *Harvard Business Review* 79 (6): 86-93.

Purdue D (2001) 'Neighbourhood Governance: Leadership, Trust and Social Capital' *Urban Studies* 38 (12): 2211-24.

Putnam RD (1993) *Making Democracy Work. Civic Traditions in Modern Italy* Princeton University Press, New Jersey.

Putnam RD (1993) 'The Prosperous Community' *The American Prospect* 4:13.

Putnam R.D (2000) *Bowling Alone. The Collapse and Revival of American Community* Simon and Schuster, New York.

Richling B (1985) 'You'd Never Starve Here: Return Migration to Rural Newfoundland' *Canadian Review of Sociology and Anthropology* 22(2): 236-49.

Schuller T, Baron S & Field J (2000) 'Social Capital: A Review and a Critique' In: Baron S, Field J & Schuller T (Eds), *Social Capital. Critical Perspectives* Oxford University Press.

Stone W (2001) *Measuring Social Capital: Towards a Theoretically Informed Measurement Framework for Researching Social Capital and Community Life* Research Paper, 24, Australian Institute of Family Studies.

Stone W & Hughes J (2001) *Sustaining Communities: An Empirical Investigation of Social Capital in Regional Australia* SEGRA 2001 5th National Conference, Townsville, 10-12 September.

Taylor M (2000) 'Communities in the Lead: Power, Organisational Capacity and Social Capital' *Urban Studies* 37 (5-6): 1019-1035.

Taylor M (2000) 'Maintaining Community Involvement in Regeneration: What are the Issues?' *Local Economy* 15 (3) 251-267.

Waddington D (2003) 'Making a Difference in Warsop Vale: The Impact of Government Regeneration Policy and Community Development on a Nottinghamshire Ex-Mining Community' *Social Policy and Society* 3 (1): 21-31.

Wainwright, S (2003) *Measuring Impact. A Guide to Resources* National Council for Voluntary Organisations, London.

Walker A & Wigfield A (2003) *Social Quality, Social Capital and Quality of Life* Discussion Paper for ENIQ, February 2003.

Wall E, Ferrazzi G & Schryer F (1998) 'Getting the Goods on Social Capital' *Rural Sociology* 63(2): 300-322.

Woolcock M (1998) 'Social Capital and Economic Development: Toward a Theoretical Synthesis and Policy Framework' *Theory and Society* 27(2): 151-208.

# 10

# Farmers' Suicides in Andhra Pradesh: The Response of Rural Political Institutions

*V. Anil Kumar*

## Introduction

In the context of wide spread crisis in the Andhra Pradesh countryside this article raises the question as to how the rural institutions are responding to the prevailing situation. The article asks the question as to whether the rural institutions should respond to the situation and why they are not responding to the situation; the paper argues that the first institutions to respond to the crisis should be the rural political institutions and not only the State level institutions; and argues that if the former are not responding this raises questions about the institutions of local governance and more so of local self governance. In the following first section deals with the notions of village community and political institutions as they are conventionally envisaged in India. The second section deals with the theoretical rationale for the user groups or what came to be known as the 'second wave of decentralization' and the impact these have had on the countryside. Thirdly, we deal with thc state of rural community and the reasons for the spate of suicides taking place in the countryside; the fourth section is a concluding section and deals with the questions as to what the rural institutions are meant to be doing in the context of the crisis discussed the preceding sections.

## Some 'Romantic' Notions of Village Community and Institutions

The present rural political institutions in the form of Panchayati Raj Institutions (PRIs) have gone through a tortuous and interesting course since their inception in 1959 (Mathew 1996). Equally interesting are the visions of village community and the notions of their governance. Here we invoke two debates about these visions and notions. And these debates have their bearing all through the rest of the paper. The first one is the debate between Gandhi and Ambedker (Omvedt 2000) on the nature of village communities and their social and political institutions. Mahatma Gandhi clearly lays down his notion of self-rule or self-governance in his classic the *Hind Swaraj*, which is a seminal text. Gandhiji also held on to a notion of village community as a village republic wherein the interdependent division of labour contributes to a self sufficient village community. While Gandhiji believed the self-sufficient village community of interdependence and social harmony as the true description of the village communities of his time, this notion has become a normative notion to build a village community on those premises for later Gandhians, Sarvodayists and followers of Gandhi. The Panchayati Raj as it was envisaged in the beginning largely followed this vision of eternally homogenous, harmonious and interdependent village communities. Ambedkerite notion of the village community on the other hand was in stark contrast to this and emphasized the iniquitous, hierarchical and oppressive nature of villages and their communities. The caste system and hierarchies are seen as the basic features of the rural communities and not the harmony and interdependence between castes. Village was a 'sink of localism' and worse. Despite the strong influence of Ambedker's critique of villages it was the Gandhian notion of village communities and their self-rule which carried the day and remained influential in shaping the policy and programmes of governance towards the Indian countryside. The Panchayati Raj institutions, however modified at present, indelibly trace their ideological lineage to Gandhian thought. We will return to this point.

The second debate that we invoke here is a more academic one that is between late Prof. M.N. Srinivas and Louis Dumont on the nature of village communities in India (Srinivas, M.N, 1996). This debate raised the question as to whether there at all existed some thing called a 'village community'. Dumont held that caste is the defining feature of the village and castes horizontally amongst villages formed objects for study rather than sociologically non-existent 'village communities'. To put in other words caste communities existed but not village as a sociological community. Prof. Srinivas sharply reacted to this position and argued that in spite of the hierarchical nature of castes and sub-castes, the village as a community, beyond the caste system, existed and therefore village as a sociological unit is a valid unit for study. And many village studies followed.

The first debate that we mentioned above is a political one and the second debate is an academic one. But both are rich debates and need consideration while discussing the state of rural communities and the catastrophes that are visiting them with such regularity. The legacy of these debates is important in the sense that the Panchayati Raj Institutions even after the 73rd Constitutional Amendment still envisage and privilege village community in the form of Grama Sabha and other related provisions. And the Gandhian vision, which is also the nationalist vision, of village communities has been considered by critiques as a 'romantic' view of the villages and is seen as unrealistic. The above two debates, that between Gandhi and Ambedker and that between Srinivas and Dumont, though have taken place at two different points in time and in different spheres still have strong relevance. While invoking these debates can we ask the basic question today, against the backdrop of pervasive crisis: do rural communities exist socially and politically? While we have touched on the aspect of village communities as social communities we have not discussed the political aspect of the village communities. Before moving on to the notions of 'user groups' we briefly dwell on the political aspect of the village communities.

There are two questions concerning the political existence of rural communities: the first one is the relationship of these communities to the larger, macro state both at national and regional levels and secondly the communities' relationship to themselves as political communities. The first question concerning the relationship of village communities to the macro state is debated since the colonial times. The colonial civil servants who chronicled their rule in India have largely described these communities as untouched by the macro state (Madan, 2002). According to this view village communities were part of larger states but were always also distant from the happenings in the macro-state. The communities paid taxes, supplied men to the armies of the larger state, but were nonetheless retained their autonomy and relative independence from the larger state. They were not fully and completely integrated into the structures and processes of the macro state. This view largely informed the earliest studies of Indian villages and has projected the picture of an unchanging, immutable village community with its own customs and traditions and systems of governance. Even Marx, informed by the writings of British civil servants in India, took to the same notion of the immutable village community; despite the view that Marx held on British rule in India, as devastating and revolutionary at the same time. This view of immutable village community is interesting and provocative even today. What is the extent of the integration of rural communities into the larger state structures? How have all the drastic changes and modifications in governance since the Independence affected the relationship between the village communities, the regional state and the national state? This question appears to be a rarely asked question and often a taken for granted question in the discourse on the rural communities and rural development. If any thing, the changes in communications, information, transport and mobility of rural populations, and the reach of the state, have largely redefined this relationship and have contributed to the present situation of the rural communities (Breman, 1997).

The second part of the question, and of more direct relevance to this paper, is the constitution of rural communities as political communities consisting of citizens participating in their own governance. As mentioned above the Gandhian notion of self – governance which informed the PRIs has undergone many changes (Mathew, 1996) and the present situation is that the communities are supposed to govern themselves within the three tier institutional framework which provides for positive discrimination and community participation (Jha and Mathur, 1999). It is thirteen years since the 73$^{rd}$ Amendment is drafted and still the conclusive evidence that the PRIs are firmly in place and are successfully carrying out the local self-governance function is a debated one. All the more so because the PRIs are envisaged to function as the third tier of governance within the polity along with the State and Union governments. Thus both the questions concerning the political constitution of the rural communities as part of the larger macro polity and as well as inwardly constituent political communities within themselves are to be posed and explored in order to be able to understand as to how these communities operate in times of crisis such as the one we are discussing here.

Taking the above discussion as backdrop and Andhra Pradesh as example we argue in the following that the rural political institutions, specifically the Panchayati Raj institutions have not been able to form or constitute a political community out of the village communities because even before the traditional village communities could constitute themselves into modern democratic political communities, even before these institutions were allowed to take root, they were swamped by a plethora of 'user groups' which fragmented and fractured the political community at the local level. The theoretical rationale and logic behind these 'user groups' is different from that of the PRIs. Below we discuss the rationale and logic of the 'user groups' that resulted in the 'second wave of decentralization'; for this logic and rationale has interesting relationship with the policy processes studies.

## The 'Rational Choice' Notion of Communities and Institutions

The creation of various user groups with connection to management of natural resources, school education, thrift and credit and so on can be seen from two stand points: one that it is expedient for government and donors to create such institutional structures. Second that though the expediency is to a large extent true there appears also to be a coherent theory behind the design of these institutions. Here we briefly dwell on the latter and return back to the impact of these institutions on village communities.

The theory that coherently explains the need and functioning of these 'user groups' is the *Institutional Rational Choice* theory and the frame work developed within the theory named 'Institutional Analysis and Development (IAD from hereon) Framework'[1] developed by Elinor Ostrom (Ostrom 1999) and her associates. This framework is an influential framework, which seeks to explain the behaviour, incentives and responsibilities of individuals particularly with reference to what she calls 'common pool resources' and common property resources in general. The framework follows a positivist approach which models upon the laboratory experiments of human behaviour and tries to extend the same positivist logic to field realities concerning the institutional structures of common pool and property resources. Besides being positivistic the framework also models the human behaviour upon neo-classical economics and presumes the human nature as that of *homo economicus*. Individuals and communities in the context of the theory appear to be utility or incentive maximizers of resource use; and the individuals' 'pay offs' and returns from the common pool resources are supposed to be proportionate to their investments in the common pool resources. All others who depend on the common pool resources appear as 'free riders'. The framework clearly explains the logic behind the institutions such as water users' associations and watershed committees and even primary schools (if we extend the analogy) all of which are seen as common pool resources. The fact that the natural resource management institutions (known as NRMIs) occupy so important a position among 'user groups' alerts us to the usefulness of this IAD framework in understanding the theoretical rationale behind the institutions and how they can be justified on theoretical grounds. The IAD framework is applied for analysis and prescription of institutional designs not only in India but worldwide.

The important point is that the 'community' as defined in this framework is the community of users of common resources but not as community of users situated in larger social and political communities. Every common- pool resource in the village can have a community of users but these users are part of a larger village and /or Panchayat community and are part of larger political structures. The problem with the IAD framework of looking at institutions becomes acute when the community of users of common pool resources build institutions based on that usage, first to exclude others and second also commanding more material, financial resources than are the other political institutions. In both terms, namely (a) excluding certain other communities who can not be exactly defined as 'users' or 'community' in the sense used in the IAD frame work and (b) when these 'user groups' or user 'communities' command large financial resources by virtue of being part of a larger polity and development strategy, in both these cases they bypass local democracy and political, democratic institutions.

Prof. James Manor in his article 'User Committees: A Potentially Damaging Second Wave of Decentralization' (Manor, 2005) argues that what is happening in the name of user committees amounts to a 'second wave of decentralization' and this wave, in contrast to the first wave of decentralization all over the world in eighties and nineties which created momentum for democratic decentralization, has damaging potential to the concept of decentralization itself. Prof Manor in a rigorous analysis of user committees with empirical information from South Asia and Africa argued that what is needed is an integration of these myriad 'single-purpose' user committees with 'multi-purpose elected councils' (Manor, James, nd) Prof. Manor also suggests that most of these user committees are donor driven, malleable to pressures from *realpolitik* and have deleterious impact on bottom-up participatory democracy. What Prof. Manor says is accurately reflected in Andhra Pradesh too as some of Prof. Manor's data is from Andhra Pradesh and some other Indian States such as Madhya Pradesh, Uttar Pradesh and Gujarat. Below we discuss the situation of PRIs and user committees in Andhra Pradesh.

## The State of Rural Institutions in Andhra Pradesh

In the past one decade Andhra Pradesh earned a reputation of being front-runner in economic and governance reforms. Given this image one would expect vibrant and well functioning rural governance in place. But contrary to such expectations the Panchayati Raj Institutions are languishing in Andhra Pradesh. Elections are conducted to these institutions but after that not much attention is paid. The Eleventh Schedule in the 73rd Constitutional Amendment requires that a total of twenty-nine subjects be devolved to Panchayats. Panchayati Raj however is a State subject and it is the will and interest of the State government to devolve functions, functionaries and finances. The previous TDP government claimed on paper to have devolved seventeen subjects. In reality except drinking water, streetlights, and internal roads and to some extent sanitation, none of the twenty-nine items are devolved. Instead, programmes such as *Janmabhoomi*, created and operated in terms of what is known as bureaucratic 'nodal system', has reversed the devolution process and concentrated powers in the line departments of State bureaucracy. This was particularly so the case with the TDP government's policy of 'de politicizing' the development. Mr. Naidu supposed to have openly said that he is in favour of development and not politics and would like to keep 'politics' out of 'development'. Following this approach many 'user committees' were created in rural areas and functioning of these was coordinated through 'nodal system' and *Janmabhoomi* Programme. Contrary to such policy pronouncements of 'depoliticizing development'[2] what took place was the packing of these 'user committees' with party cadre and using the user committees for party building (Powis, Benjamin, 2003). Interestingly this policy was in contrast to both the early record of Telugu Desham Party which abolished hereditary village officers, created smaller Mandals out of bigger Taluks as sub-district, intermediate tier of governance and projected an image of a populist political party. Secondly, this policy was also in contravention to the liberalizing thrust of the government. The liberalization, privatization and globalization set of reforms are closely associated with the idea of drive towards decentralization and moving away from centralization of power. In Andhra Pradesh, on the contrary the reforms were associated with centralization of power in the bureaucracy and at the State level. This policy trend has marginalized Panchayats. In addition to this the policy environment over the past decade was also markedly hostile towards attempts of the civil society to rejuvenate PRIs and strengthen local self-governance. Thus reforms and liberalization in Andhra Pradesh have taken place reversing the trend of decentralization; concentrating decision making authority in bureaucracy and discouraging civil society from actively pursuing the strengthening of Panchayati Raj Institutions. The focus to a large extent was on 'user committees' that were created at the village level. Andhra Pradesh witnessed 'second wave of decentralization' before people at village level could fully comprehend the significance of 'first wave of decentralization'. We discuss some of its details below.

The situation in Andhra Pradesh is compounded, with the introduction of user committees at the village level. Some of them are statutorily introduced and others through executive fiat. The Water Users Association, the Watershed Committee, the Parents and Teachers Committee the Vana Samrakshana Samiti the Mothers' committee and added to this are the numerous self-help thrift and micro-credit groups. The institutional landscape in the countryside is certainly complex and this is far from being advantageous. The complexity has become a disadvantage in mobilizing the community in times of crisis. This multiplicity of institutions has actually contributed to the fragmentation of the social, and more so, political community.

Unlike the expectations, these user committees have also acquired the negative attributes of the representative institutions at other levels. Money and muscle power are used in their elections; liquor and meat are generously provided to the clientele to secure success in the elections. And the victory in water user committee election or the parents and teachers' committee election has political rewards. As for the aspects such as protective discrimination and gender equality go even in these institutions the posts reserved for women are dominated by men and where Dalits or Backward class candidates are supposed to occupy the positions it is the upper caste individuals who dominate the institutions either by proxy or by other means.

Another important aspect of these parallel institutions to PRIs is that, as Prof. Manor suggests, most of the time these are created with the support of, and following the dictates of, the donors who provide funds to these institutions. For example, in Andhra Pradesh, Parents and Teachers Committees are created through an organization called DPEP (District Primary Education Programme) and are supported by the DFID. Like wise the Watershed Development Committees are created and run with the support of DFID. Water Users Associations and Vana Samrakshana Samitis (VSS) are supported by the World Bank. The argument that while PRIs provide opportunity for people to participate in decision making process in one context while the user committees provide opportunities for participation in other more specific contexts does not seem to explain the problem because both user committees and PRIs operate in the same social and political context.

The major problem with the institutional landscape described above is that there is lack of horizontal connection between any of these institutions. It is the contention of this paper that these institutions being disparate and distinct do not help foster social solidarity or social capital. These bodies lack horizontal connections with each other and also lack connection with the panchayats; most of the time the user committees are under the control of local bureaucracy. What the prevalence of this multiplicity of user groups has done is to take away some crucial functions from PRIs, which would otherwise provide PRIs with authority, legitimacy and resources. This aspect, as Prof. Manor notes, creates confusion among the ordinary citizens about the genuine source of leadership for the community. Some of the functionaries of these institutions are in practice more powerful and generate more patronage than PRI leaders because they are better funded for one single aspect e.g. watershed, while Panchayats as 'multiple use councils' are less funded.

What implication does this institutional landscape hold for rural governance? The governance of water resources becomes separated from that of watershed; Anganwadi from that of the school; watershed from that of forest committee; and local Panchayat from all these. This is a confusing landscape of institutions for an ordinary farmer. The pertinent question in this context is what is the one institution that listens to the plight of the farmer? Certainly the votaries of the user committees—that are also known as 'CBOs' i.e., Community

Based Organizations—appear to claim that it is none of the tasks of any of these institutions.

The point that is considered is that in the context of the above governance at the local level has become no body's task. The political institutions and the individuals that occupy them do not see that the minimum welfare of their citizens is their concern. The leaders of these institutions do not think that the tragic state of affairs has to be dealt with by coming together, by representing, lobbying, putting pressure, protesting or by agitating. None of the *Sarpanches* in any of the villages, where suicides have occurred, ever seem to have felt this. Sarpanches association of Andhra Pradesh or even ZP chairpersons' association has not bothered themselves to make any statements about the agrarian distress or suicides. What is of concern is that not only that empirically they have not done anything but the normative standards of the larger society and polity which do not expect the rural institutions that are in the immediate vicinity of the farmers to respond.

This situation is particularly so with the case of Andhra Pradesh. If thousands of farmers are committing suicides, if regular reports of suicides are coming from all parts of the state then why the rural institutions are are not responding to these occurrences? Are they not responsible for governance of rural areas?

## The State of Rural Community and the Reasons for the Suicides

As it comes out from the reports of the farmers' suicides it is economic, ecological and social reasons, which are driving farmers to suicides. The economic causes for suicides as are reported: the faulty inputs such as spurious seeds, pesticides and fertilizers, their high prices; the low and widely fluctuating prices and inadequate marketability of the output; the failure of yields and lastly the only economic *institution* that is cited as reason is failure of the institution of credit. Private moneylenders and non-institutional credit is often the reason for suicides. The other less cited institutions are the leasing in and out practices. These are the main economic reasons. The iniquitous distribution of land forms the agrarian backdrop. Often these reasons can be traced to the macro-economic policies of the Indian state, which has been neglecting agriculture. That the state intervention in promoting primary education, primary health and land reforms has always been minimalist is the argument forcefully made by Amartya Sen and Jean Dreze (Dreze Sen, 2002). While the state presence in these aspects has been relatively less, the neglect of agriculture and allied sectors in Andhra Pradesh was particularly glaring. In Andhra Pradesh the State was increasingly becoming minimalist in some aspects such as agricultural extension and animal husbandry aspects in which the state machinery did provide some support in earlier period of time till mid eighties.

The ecological reasons are drought, the fluctuating yields associated with drought and the devastation caused by the natural-scientific establishment, or the façade of it, that is associated with the intensive farming. Mono-cropping associated with the green revolution technology is frequently cited as one among the major reasons for the crisis. The shift from multiple-cropping and food economy to mono-cropping particularly towards the crops such as cotton and chilly is the major reason for the vulnerability of the farmers. Here the one institution that comes under scrutiny is the Agricultural Extension Department that has been invariably a failure. Increasing desertification of parts of Andhra Pradesh is also a reason for falling levels of ground water and creation of a hydrological crisis as a sub-set of ecological crisis. This set of ecological reasons can also be traced to the gradual withdrawal of the state from irrigation, extension and regulation.

The less asked reasons are the social reasons. How is that these individuals who are driven to commit suicides are not prevented by the rural community. Not supported economically or socially or psychologically by the members of their community? Here we are sure that the 'community' in the countryside is caste divided and class divided and is also gender divided. Sociologists, as discussed in the beginning, have always held that the rural communities are caste divided and further divided into sub castes; but they have also informed us that there are institutions of caste endogamy, commensality, kinship, consanguinity, caste panchayats and caste associations. Where have all these social institutions gone in the face of the dilemmas facing the farmers? Have the communities with their social ties dissolved? Has development of market economy in the rural areas with its social logic dissolved the community relations and supports? Has individualization proceeded thus far? It is often observable that the farmers are not facing the crisis collectively in organized protests but are only committing suicides. Often these questions as to what happened to the community ties, however primordial one would like to describe them from modern point of view, are the less asked ones. The following section discusses these questions with reference to brief fieldwork conducted in Warangal District of Andhra Pradesh.

## A Field Report from Atmakur

Fieldwork was undertaken in Warangal district of Andhra Pradesh keeping the above realities in mind. Warangal district is most affected by the suicide epidemic and suicides have been taking place in the district since 1997. The district is most advanced district in Telengana in terms of agricultural practices and commercial crops such as cotton and chilly are grown regularly on its fertile black soils. The district also fares relatively better in terms of rainfall compared to other districts of Telengana and therefore makes interesting case for study. Within Warangal we have chosen Atmakur Mandal for fieldwork. Among all the mandals that are affected by suicides Atmakur mandal is worst affected and this made us chose for fieldwork. Atmakur mandal headquarter i.e., the Atmakur village itself has seen as many as nine suicides of farmers. We have interviewed five cases out of these nine in addition to interviews with the Sarpanch, the mandal Agricultural Officer (AO) and others. Firstly we briefly discuss the main highlights of each of the five case studies of suicides and then discuss some common points that emerge.

### *Case Study –I*

Basboina Sudhakar committed suicide on $28^{th}$ November 2002. Sudhakar was a farmer owning two acres of land and was from Mudiraj community. Mudiraj community is a backward class community. He owned these two acres and cultivated rice on the land with water for the crop being pumped from a nearby stream. But two years prior to his death he has leased in additional three acres from another backward class person and cultivated cotton crop. He died owing to the burden of accumulated debts owing to the cultivation of cotton crop and failure of paddy crop. On the eve of his death he cultivated paddy on his two acres of plot and cotton on the three acres of leased land. The lease per acre per year was Rs. 5,000. Prior to shifting to cotton crop on leased land, he was a rice farmer. Interview with his family members show that he accumulated debts owing to the failure of cotton crop and drying up of water for paddy crop. As for the cotton crop the seeds, fertilizers and pesticides have all failed. Recurring failures of rice owing to drought and of cotton owing to faulty inputs has led to the suicide. The accumulated debts were about 1, 50,000 Rupees. Sudhakar borrowed all this money from his close relatives of the same caste.

There are about 200 families of the same caste and all are marginal and small farmers and they also conduct petty trade. After the suicide the family did not have money for the funeral expenses and the caste community bore that expenditure. The close relatives who lent Sudhakar the money however expect the money to be paid back to them. Sudhakar himself never shared his psychological pressure and situation with his wife and family members when he was alive.

After the suicide the family sought help from the local Panchayat and Mandal Revenue Office for compensation or exgratia. So far they have received none. The officials have taken note of all the details of the occurrence and promised support. The only support the family received so far is twenty kilos of rice given under food for work programme in two installments. The village has a Grama Panchayat, a Water Users Association and a Parents and Teachers Association. These institutions, however, have been of little help. The caste community now provides social and psychological support for the family. The Mudiraj community still has practices of caste endogamy, commensality, kinship and a caste Panchayat that settles disputes within the community. Sudhakar's family members inform us that the caste community did not matter when the economic crisis came.

Jannarapu Komuraiah suicided on 14th December 1999. He owned one and half acre of dry land and cultivated two acres of leased in land. He leased land from a local land owner from Reddy caste at the rate of Rupees 2000 per acre/per year. Komuraiah was a Scheduled Caste farmer. He cultivated cotton crop on the total three and half acres of land. He was cultivating cotton for past eight to nine years prior to his death. Komuraiah died owing to debt burden that accumulated owing to cultivating cotton. He owed around Rs. 80,000 at the time of his death. The cotton crop failed continuously for two years prior to his death: first year owing to excessive rains and the consecutive year owing to drought. Komuraiah borrowed all the money from his relatives both in this village and in neighboring villages. The lenders never pressurized him to pay back but he himself went into depression owing to debt and committed suicide by consuming pesticide.Local institutions were of no help any time. So far Komuraiah's family has not got any compensation or exgratia. The Mandal officials and the Grama Panchayat have promised exgratia and the family is waiting.

Manda Sammaiah died consuming pesticide on 17th December ,2001. He owned two acres of dry land and leased in another two acres from an upper caste landowner at the rate of Rs. 3,000 per acre/per year. Sammiah was a Scheduled Caste farmer who cultivated cotton on his own and the leased land. His family along with the other caste members lives in the caste-demarcated street till now. At the time of his death Sammiah had debts amounting to Rs. 1, 00,000. Of this, 40, 000 Rupees were borrowed from the same land owner who leased out the land at four percent interest. And the remaining amount was borrowed from his relatives. Those who lent money pressurized Sammiah to pay back; and under the pressure he suicided. Exgratia was promised and the family is waiting .The Mandal Development Officer and Mandal Revenue Officer say the compensation will come soon.

Parikirala Sadaiah died of pesticide consumption on 16th September 2000. He was a backward class farmer owning three acres of dry land and 0.75 acres of wetland. Sadaiah died owing to accumulated debts of about Rs. 1,15,000. The debts accumulated because of the failure of the cotton and chilly crops and owing to a failed attempt to dig a well in his dry land. The well failed owing to a rock layer that came after digging into some depth. The interest on the loans has multiplied. Sadaiah brought money for his crops and well mostly from his relatives. And the lenders pressurized and humiliated.

The major reason for the crop failures was failures of the cotton seeds to germinate and because of the pesticides that did not work on the pests. As his family members say seeds failed, fertilizer failed and pesticides failed. There was no help from any side. Local Sarpanch and other village elders contributed to the funeral expenses.

There are caste related practices and institutions such as caste Panchayat, endogamy and commensality. But the members of the caste were not of much help be cause they are also small and marginal farmers and could not have afforded to help Sadaiah in paying his debts. The caste community has a caste chit fund and Sadaiah borrowed some money from the chit fund too. That money too is to be paid back. There is no exgratia or compensation from the government so far. The family still cultivates cotton; apparently owing to the reason that the crop withstands drought condition better which the other crops do not.

Reguri Gopal Reddy died on 17th February 2000 consuming pesticide. Gopal Reddy had land about two acres and leased in two acres of land. He cultivated cotton on four acres. He died of pressures from accumulated debts and other losses owing to digging of a well and death of plough animal. The relatives of Gopal Reddy are well off farmers and do not care about the situation of his family. The wife of Gopal Reddy now gets the field cultivated by employing an annual farm servant. The family still cultivates cotton crop. No exgratia or compensation of any kind reached the family so far.

The five case studies that we discussed above share some common points. They are:

- All the five farmers are small/marginal farmers cultivating two to five acres and almost all except one have leased in land in order to cultivate cotton; their caste composition is that two are from backward class; two are from Scheduled Castes and one farmer is an upper caste person.
- All the farmers are cotton farmers and died because of accumulated debts owing to cotton crop failure, drought and failure of wells, and some contingent factors. The villains of the piece prominently were the cotton crop and drought.
- One interesting feature is that all the farmers discussed above have borrowed money from their close relatives and caste members. Caste solidarity did work in getting loans but failed when the paying back could not result.
- One important observation that is discernible from the case studies is that agricultural extension has totally failed as far as the crops go and none of the farmers whose cases we discussed above availed any extension services. None of them knew what seeds, pesticides, weedicides and fertilizers should be used, when, in what quantity and how. There is also a big question mark on the quality of the all of above inputs on which there was no advice.
- The last and most significant observation that emerges is that in none of the cases did the local institutions matter in any way. Not even to expedite the reach of compensation or exgratia.

The discussion with the *Sarpanch* of these cases also did not yield much in terms of the understanding of the relationship that the Panchayats can play in such circumstances. He said that he has not convened a Grama Sabha to discuss the situation but discussed the farmers' distress when the mandatory Grama Sabha was conducted. The Sarpanch has issued a broad appeal to the farmers that they should not commit suicides when faced with adverse agricultural conditions. The Sarpanch however contended two points: one the user committees have fragmented the political community and that these user committees have largely been floated to strengthen the cadre of a particular political party and secondly, that the requisite awareness to come together, discuss a way out of the impasse does not exist among the farmers and villagers in general. According to him that process may take time. At present, the Sarpanch claimed, the Grama Sabha attendance is only 100 persons out of 9,000 odd voters in the Panchayat.

The case studies presented above also show that agricultural extension as an important aspect of intensive agriculture has failed. Following this observation we conducted detailed discussion with the Agricultural Officer of Atmakur Mandal, which resulted in some interesting insights the main points of which are presented below.

- One of the main reasons for the agrarian distress is the increasing adverse seasonal conditions such as droughts and excessive rainfall and these conditions are particularly acute since 1994-1995. In the presence of favourable seasonal conditions the extension services, as they exist now, can help attain good return from the crops with ninety percent assurance. But in the presence of adverse seasonal conditions extension services can only recover the investment with great difficulty.
- As the agricultural conditions exist cotton is the only crop, which can give minimum guaranty of crop return and monetary returns. That is what makes farmers to stick to this crop and farmers are continuing this crop even after suicides are taking place in their families owing to the failure of the same crop. Agricultural scientists have suggested to extension workers that they should advise farmers to shift to crops such as soya bean instead of cotton but soya bean also can not withstand drought conditions when compared to cotton.
- Presently the extension services are working with meager staff. Presently one Agricultural Officer and two extension

workers cover twenty seven Gram Panchayats in the Mandal and there is virtually no field staff or infrastructure to reach out to the farmers. Present strength of extension staff is grossly inadequate. The policy of employing multi-purpose extension officers (MPEOs) at the rate of one worker per every 1000 hectares of cultivated land in the Mandal needs to be implemented. In the absence of field staff the agricultural officer at the mandal level is only running a 'one man show'.

- The spurious seeds and other inputs is a major problem. To curtail impurity in seeds there should be a separate 'seed inspector' at a mandal or division (cluster of mandals) level'. This is important because faulty seeds cause loss of entire crop in a season. Monitoring of seed supply needs to be strengthened.
- At present the seed producers and marketers themselves do not know the plant characteristics and duration of the resultant crop etc. These people influence farmers to purchase by informing whatever they think is correct. Often both the buyer and seller both do not know any thing about the seed/pesticides/inputs they are dealing in.[3]
- As the situation exists now details of seventy percent of the seeds and pesticides, as they are available in the market, are not known to the extension department officials. Sale and purchase of the seventy percent of the seeds and pesticides and fertilizers takes place out side the purview of the extension department. Therefore the technical aspects of whatever seeds, pesticides newly enter into the market should be made available to the extension staff.
- At present the extension department at Mandal level lacks both field staff below to operate and scientific guidance from above. And there is often no coordination between Panchayats and existing extension staff.

The details of suicides, the views of elected representative and the situation of the extension department as they characterize the crisis are presented above. Role of these institutions has been crucial to the complex situation.

## Conclusion

In the foregoing we have discussed the notions of village community and institutions; the reasons for suicides and the role and the situation of the village level institutions. It comes out clearly that village and village community as social and political communities do not matter as far as addressing the conditions that drive farmers to commit suicides are concerned; and so is governance at that level. The faith in PRIs as foremost institutions for rural governance has been placed by the 73rd Constitutional Amendment because they are the immediate institutions to the local social and political communities consisting of citizens. While it is obvious that these institutions are not able to take up the issues at the local, village level, leave alone at any higher level, what is lamentable is not only the empirical fact that they are not addressing themselves to this issue but the fact of normative standards in the discourse about institutions that do not raise this question. Even in the neo-liberal paradigm of development and governance, it is the triad of market, civil society and the decentralized governance that are supposed to take care of the economic, social and political welfare of the local communities. The situation of routine occurrence of the suicides indicates that while the older paradigm of centralized state has been discontinued that has not been replaced by efficient and accountable market, strong and well developed civil society and effective, decentralized local self-governance that cares about the welfare of its communities .The interesting point that emerges is also that while the Panchayats as political communities continue to be a) loosely integrated into the larger regional and central state structures b)as political communities constituting the third tier of governance they are still not fully 'mature' to take decisions collectively to address the problems facing its citizens. The idea that the local self-government, in the form of PRIs, will lead to decision making on the part of the citizens to arrive at decisions that affect their lives is still, a far cry. Sociologically the communities are not totally dissolved and caste loyalties to some extent are present. The case studies show that in all the cases discussed the farmers have mobilized loans initially from their relatives from within the village or out side. But these identities and loyalties are in a transition and the force of the market is increasingly individualizing the farmers. The situation is such that while the traditional community ties, which Gandhians admire so much, are giving way, new modern political communities where the citizens are aware of their citizenship and exercise that citizenship to take decisions that affect their lives is still not born. It is difficult to write conclusions in such situations on a democracy and self-governance that is only thirteen years young. Whether the local communities assert themselves to address and solve their problems at that level is a question only time can answer.

## NOTES

1. The framework distinguishes between two notions of institutions: one that is a structure and second that is a routinised practice. Though the IAD framework as explained by Elinor Ostrom emphasizes the latter, the entire discussion collapses both meanings and connotes more often the former.
2. Prof. James Manor calls it an important justification given for 'user committees' and calls such an attempt at separating politics and development 'a myth'.
3. These aspects clearly show the reasons for calling the natural –scientific knowledge regime, which is operating in agriculture a façade, and has contributed to the distress.

## REFERENCES

Aziz, Abdul and David D. Arnold (1996). *Decentralised Governance in Asian Countries.* New Delhi: Sage Publications.

Breman, Jan (1997). "The Village in Focus" In: *The Village in Asia Revisited*, Jan Breman, Peter Kloos and Ashwani Saith eds.), Delhi: Oxford University Press, pp,15-75.

*Down to Earth* (2004). "Andhra Agriculture: Suicidal state". (*Cover story*), 13(.4), July 15., pp.15-24.

Dreze, Jean and Amartya Sen (2002). *India: Development and Participation.* Delhi: Oxford University Press.

*Frontline* (2004). "Death Trap." (Cover story), 21(13), June 19 - July 2.pp.4-29.

Hanumantha Rao, C.H and S. Mahendra Dev (eds) (2003). *Andhra Pradesh Development: Economic Reforms and Challenges Ahead*. Hyderabad: CESS.

Haragopal, G and G. Sudarshanam, (2000). "Andhra Pradesh" in *Status of Panchayati Raj in States and Union Territories of India*. New Delhi: Institute of Social Sciences, pp.45-58.

Jain, L.C. (nd). *Cry the Beloved Self Government*, New Delhi: PRIA.

Jha, S.N.and Mathur P.C. (ed.) (1999). *Decentralization and Local Politics*. New Delhi: Sage.

Leftwich, Adrian (1990). "Politics and Development Studies" in *New Developments in Political Science: An International Review of Achievements and Prospects*, ( Adrian Leftwich ed.) Aldershot: Edward Elgar.

Madan, Vandana (ed) (2002). *The Village in India*. Delhi: Oxford University Press.

Madhavan, Samiksha, Ramana (ed.) (nd), *Farmers' Suicides in Andhra Pradesh*, Hyderabad: AWARE.

Mahendra Dev, S (2004), *Is Rural India Shining*? CESS Working Paper: 54, Hyderabad: CESS.

Manor, James (1996). *The Political Economy of Decentralization*. (Mimeo)Sussex: Institute of Development Studies University of Sussex.

Manor, James (2005) "User Committees: A Potentially Damaging Second Wave of Decentralization" (mimeo) Sussex: Institute of Development Studies, University of Sussex.

Mathew, George (1997). "Panchayats: The Local Self-Governance System in India" In: *Decentralised Governance in Asian Countries*, Abdul Aziz and David D. Arnold (eds.) New Delhi: Sage Publications, pp. 125-145.

Mitra, Subrata, K (2001), "Making local government work: local elites, *Panchayati raj* and governance in India" In: *The Success of India's Democracy*, Atul Kohli (ed), Cambridge: Cambridge University Press.pp.103-126.

Omvedt, Gail (2000) "Gandhi and Ambedkar: Why the Confrontation?" in *B.R. Ambedkar: Life, Work and Relevance* M.L. Ranga (ed), Delhi: Manohar Publications, pp.83-89.

Ostrom, Elinor (1999). "Institutional Rational Choice: An Assessment of the Institutional Analysis and Development Framework" In:

*Theories of the Policy Process*, Paul A.Sabatier (ed) Boulder: Westview Press, pp 35-71.

Powis, Benjamin. (2003)."Grassroots Politics and 'Second Wave of Decentralization' in Andhra Pradesh", *Economic and Political Weekly*, June28-July 4.

Reddy, Gopinath, M. (2003). "Decentralised Local Bodies: Post 73rd Amendment Scenario" In: *Andhra Pradesh Development: Economic Reforms and Challenges Ahead,* C.H. Hanumantha Rao and S. Mahendra Dev (eds), Hyderabad: CESS, pp. 599-623.

Srinivas, M.N.(1996). *Village, Caste, Gender and Method: Essays in Indian Social Anthropology*. Delhi: Oxford University Press.

Srinivasulu, K (2004), *Political Articulation and Policy Discourse in the 2004 Election in Andhra Pradesh,* GAPS Working Paper: 1, Hyderabad: CESS.

Turner,Mark and Hulme,David (1997). *Governance, Administration and Development: Making the State Work.* Connecticut: Kumarian Press.

Vaddiraju, Anil (2001). "Panchayati Raj Institutions and *Janmabhoomi* in Andhra Pradesh", *Participation and Governance,* 7(21) pp.6-9.

Vaddiraju, Anil, Cheruku Venkatesham and Kodali Savitha "Primary Formal Education and Panchayats - Experience of Andhra Pradesh", *Participation & Governance*, 8(25) pp.28-30.

Vyasulu, Vinod (2003). "Rural and Other Development Projects: The Question of Institutions", In: *Panchayats, Democracy and Development*, Vinod Vyasulu (ed), Jaipur: Rawat Publications, pp.18-45.

# 11

# Community-based Disaster Management : Formulating a Community Safety Plan to Tackle Disasters

*Ranabir Dey*

## ABSTRACT

Environmental hazards may be defined as those extreme events either natural or man-induced, which exceed the tolerable magnitude within or beyond certain time limits, make adjustment difficult, result in catastrophic losses of property, income and lives and become headlines of different news media at world level. Natural environment hazards involve comparatively rare high density processes and extreme events caused by both terrestrial and atmospheric processes. Where a hazard impact is such that normal community and organizational arrangements are no longer functioning, disaster can be said to have occurred. Natural disasters take a huge toll of human lives and property and severely hamper development. It may not be possible to prevent natural disasters totally, but with careful pre-disaster planning and preparedness, we can reduce the adverse effect of the hazards and the economic loss. Here evolves the significance of disaster management. Disaster Management means a planned and systematic approach towards understanding and solving problems in the wake of disasters. It involves the systematic observation and analysis of measures relating to disaster prevention, mitigation, preparedness, emergency response, rehabilitation and reconstruction. It may be realized that disaster management is a function of community preparedness. A natural hazard can transform into a natural disaster depending on its impact on society in terms of loss of life and property. In order to enhance community preparedness, a proper safety plan is what is essential. The community preparedness plan involves all pre-disaster planning to reduce the loss as a result of natural disaster. It is basically a synthesis of various specific plans to solve a common purpose. This paper presents the various methods to be followed in the formulation, development and implementation of a community safety plan that forms the backbone of community based disaster management.

## Introduction

Disaster is an event of natural or man-made causes that leads to sudden disruption of normalcy within society, causing widespread damage to life and property to such an extent that normal social and economic mechanisms become inadequate to restore situation. Natural Disasters are a common occurrence now a days. The devastating floods, cyclones, earthquakes, droughts and other disasters are merciless in respect to the immense damages that they cause in terms of damages that they cause in terms of loss of human life, property and sanity, above all. The irreparable damages often seriously affect the progress of the country, as a whole. The extent of disaster depends upon the nature of the hazard (physical severity, frequency of occurrence) and vulnerability of the community. Natural disasters disrupt normal life and environment, account for huge loss of human lives and livestock and causes economic loss to the nation by destruction of buildings, human settlements and infrastructure, dislocation of transport and communication systems etc. The immediate effects of disaster provides for a recurring impacts like health hazards, decreased agricultural production. It is basically because of the community remains unprepared, exposed and unaware and therefore unable to assess warning systems effectively.

The need for an effective disaster management strategy to lessen disaster impact was increasingly felt. Proper planning and effective preparedness like identification of hazards relevant to a country, use of appropriate measures for mitigating the impacts of disasters can reduce the nature and extent of economic loss and adversities that are related to natural disasters. Management of natural disaster is a multi-disciplinary, inter-related and co-ordinated framework which requires involvement of experts from various disciplines as well as governmental, non-governmental and voluntary agencies coming up with their co-operation. Although disasters are active since time immemorial yet prevention and preparedness measures are of recent origin. *Disaster Management* means a planned and systematic approach towards understanding and solving problems in the wake of disasters. The science of disaster management involves the systematic analysis of measures relating to disaster prevention, mitigation, preparedness, emergency response, rehabilitation and reconstruction. There had been a growing shift in approach to disaster management, wherein the emphasis which had been to relief earlier had now changed towards prevention and preparedness aspects to reduce the damaging impact of natural disasters.

Disaster prevention and preparedness consists of a variety of measures ranging from long term to short-term as per their effectiveness are concerned. Prevention is more concerned with generating long term policies and programs to prevent or eliminate the occurrence of disasters while preparedness relates to various forms of operational planning, education and training that will prepare the population of disaster prone areas with respect to relief operation and measures. However, an interrelated framework is desirable and essential between the two to maximize the effectiveness and harmony. Proper and close collaboration of a diversity of people and organization, allocation of wide ranging tasks and development of skills for different persons and a carefully structured training program are inevitable to achieve the desired goals in disaster management.

However, in spite of implementing all the above mentioned procedures during the project period, it had been observed that several disaster management programs have failed to achieve a level of sustainability at the local level. This is mainly because several such programs lacked proper and effective community participation. The community experiences the first hand vigorous and disrupting impacts of natural disasters and therefore their participation in all aspects of disaster preparedness—immediately before, during and after the occurrence of the event—becomes a fundamental and also the most important element of any such disaster management programs. The most common forms of community involvement are partnership, participation, empowerment, and ownership by the local people. The main emphasis of disaster management should definitely be on communities and people who live in them. Unless disaster management efforts are not sustainable at the individual and community level, it is difficult to reduce the losses and scale of the disaster. Every programme or project should provide an opportunity to the local people to effectively participate in decision-making as well as in the implementation of such measures. Involvement of communities is important in both pre-disaster planning and post-disaster restoration and recovery process.

## Community-based Disaster Management

It had been evident that several disaster response programs and framework had failed to reach their desired results due to their approach towards the disaster management. The approach more often than not had been a top-down approach and a logistic-centered approach which fails to incorporate community participation at different levels of plan formulation and implementation. These reasons coincide to act as a hindrance for appropriate level of meeting and discussions, which in turn prevent any such programs to achieve the sighted needs. Use of external resources, and an indifference towards local factors acts as a barrier between disaster management plans and programs and their optimum utilization. Keeping in mind these facets and also the limitations, the *Community-based Disaster Management* approach is to be adopted. The community-based Disaster Management approach nullifies the concept of general dissatisfaction as it is basically involves modeling a *bottoms-up (community and local level)* approach towards an integrated responsive disaster management system. The community-based disaster management approach rectifies the top-down approach wherein a major shift from emergency management of risks to disaster risk management had been adopted.

Besides, the community-based disaster management approach also highlights and emphasizes the importance of community involvement in the formulation of community safety plan and also in its implementation. After all, people in the community are the worst sufferers who actually face the disaster damage and consequences in the very first hand. To define, the community-based Disaster Management approach involves activities, measures, projects and programs to reduce disaster risks to people living in high-risk communities, and are based on their urgent needs and capabilities. To be effective, local communities must be involved in identification of hazardous conditions, their proneness towards such conditions and their capabilities to fight against them. The community-based disaster management approach enhances the ability of the people to respond effectively and spontaneously to emergencies and awareness regarding control over resources and basic social services.

To distinguish traditional approach with Community based disaster management approach, the main difference lies in the general outlook towards disasters. The former basically looks into disasters as unforeseen and unavoidable events, whereas the latter convinces disaster avoidance through preparation, which can reduce damages if not prevent the event altogether. Traditional approach focuses upon emergency response and recovery while community-based disaster management approach puts stress on disaster management through measures of prevention, mitigation and preparedness. External aid is the only remedy for the affected people under traditional approach while community-based disaster management approach strengthens people to actively participate in recovery acts. People's existing capacities are used and strengthened. The community-based disaster management approach allows people to analyze the existing situation on the basis of their past experiences. Not only community-based disaster management provides emphasis on public participation in both decision-making and implementation of plans and programs, but also it also realizes the importance of scientific and objective risk assessment and planning. In this context, it should be kept in mind that *risk management*, which includes identifying threats, determining their probabilities, estimating their impacts and identifying reducing mechanisms to minimize threat, forms an integral part of disaster management.

The community-based disaster management approach put forward the concept of community-based disaster management where the importance of communities are underlined. Public participation, identification of the most vulnerable section of the society, assessment of existing situations and their standards, and implementation of risk reduction measures are the major components in community-based disaster management. Community based disaster management mainly aims at reducing vulnerabilities and increase in capacities of the community involved. The goal is to prepare a safe, disaster resilient and developed community and also add sustainability to the measures that has been undertaken. Technical training of masons and engineers are provided to ensure the quality of services, disaster management planning and training is also provided for local governments, community leaders and NGOs to ensure efficiency in the services are the major component of any training module of community-based disaster management. The basic objective is to develop a training module which would utilize local materials and local materials and local initiatives to set up a *community safety plan* to prevent disaster henceforth.

The community-based disaster management approach realizes the importance of as many stakeholders involved in the process, as this would ensure the desired goals to be achieved more rapidly and effectively. Transferring the resources to the local people and extracting local benefits from there which finally gets utilized in community development itself remains a top draw in every feature of community-based disaster management, which focuses towards disaster reduction as a whole.

## Community Preparedness Plan

In view of the fact that the natural disasters vary both in nature and magnitude in different countries, a country should prepare a community safety plan which has a regional suitability and be effective in mitigating the disaster risks. The community preparedness plan involves all pre-disaster planning to reduce the loss as a result of natural disaster. It is basically a synthesis of various specific plans having a common focus. The plan formulation is the duty of a safety committee, which gets appointed formally. The major aim to incorporate the strategies of risk reduction into future community safety plans. The various aspects of community preparedness plan are as follows :

- ***Identification and assessment of the hazard***—A general conclusion is reached regarding the natural hazard only after several literature reviews, collection of data and interviews. Besides, a close scrutiny of previous planned developments regarding disaster preparedness also forms an integral part, which would highlight the pattern in which such atrocities can be avoided. Several such informations would make easily accessible for the public by the review committee.
- ***Safety Goals and Objectives*** — Once the hazards have been outlined, the first and foremost thing that comes up is to quantify the amount of losses inflicted by the disaster. However, this very process seems to be quite critical because of absence of a benchmark, in this field. But the death caused by disasters has been made mandatory instrument in loss quantification. Ironically, it is quite difficult to measure high mortality from rare events such as tornadoes, earthquakes or flash floods. For this reason, it is important that simulations of potential mortality from such disaster agents be carried out before strategies to meet mortality goals are finalized. In addition to the fatal incidents, enormous financial costs are also incurred, which the safety goals aim to reduce. Records of insurance companies, police and government department of disaster mitigation, economic analysis, urban affairs, transport and agriculture.

Once such estimation regarding the losses has been made, it is followed by the process of establishing the safety goals. Deciding upon the safety levels is a nerve breaking exercise and it will inevitably invite several conflicts of ideas, opinions and judgments. However, a common conclusion dealing with the acceptability of the risks is achieved, only after analyzing various approaches such as *risk aversion*, which primarily focuses on risk reduction regardless of the costs and is not popular among involved community preparedness planners, *risk balancing*, which considers risks as socially acceptable only till certain extent, which infact is determined only after consultation of appropriate reference cases, such as technologies, natural background levels or risks from other activities and comparing them, *cost effectiveness,* which tries to maximize the benefits available from the funds allocated and spend on safety measures, *cost-benefit balancing* approach recognizes that some level of risk can be acceptable if other objectives are to be met. It is decided by balancing the benefits of an activity against the level of risk it entails.

## Potential Strategies

Once the level of losses have been estimated due to the natural disasters, a strong step is essential to set goals for their reduction. To begin with, a proper understanding of the reasons for their occurrence is important to be analyzed. This knowledge would enhance the level of action in designing strategies for land use, operating procedures and design criteria, which would ensure proper fulfilling of the targets. Three basic steps are essential for framing any strategy regarding disaster mitigation, namely, *remove the hazard, avoid its consequences, or mitigate its impact.* Each alternative strategy to minimize social and economic losses differs in terms of cost, timing, equity and distribution of effect.

The first step in formulating safety strategies is *to identify the major hazards in the region.* Once such informations are collected, various mechanisms could be put to use. Based on them, the risk factor involved in the disaster such as *mapping and risk micro-zoning* of the hazard prone areas, *land use planning* of future development activities including location of settlements and other development projects e.g. dams, industrial complexes, power stations and townships can be made. It is also the responsibility of the safety committee to actively participate in the evaluation of new developmental activities. Necessary land use legislations and regulatory measures are implemented to restrict or prevent industrial and residential development in high risk potential areas. In this way a variety of precautionary measures for reducing losses from particular hazards can be identified. However, it should be kept in mind, that the final decision of selection and implementation of a particular strategy lies with the political bodies, and therefore the safety first principle would be primary goal rather than economic and environmental benefits. Therefore, a major emphasis on land use regulations and zoning ordinances on the basis of cost effectiveness, administrative transparency and dearth of environmental impact is inevitable.

However, the choice of potential strategies is not a one-way traffic. It involves various held-backs. Seven such constraints have been identified by Fischhoff *et al.* (1978). They are *lack of sufficient knowledge to link causes with effects, an unwillingness to forego associated benefits, a limited capacity to react, divergence between perceived and actual risks, conflicting goals, institutional weakness and the potential for creating new hazards.* Society has multiple goals and the need for safety must complete with others such as desire for employment, food, water, transportation and energy. In some instances, the national or local goals may clash with the safety measures.

## Safety Programmes

Once a safety committee reaches a conclusion regarding the nature and extent of a disaster affecting the community, several methods are adopted to shrink the risks and the consecutive losses. Formulation of safety programmes is one such step. They involve determining the spatial distribution of risk, increasing safety through better design, disaster prediction, establishing warning systems and the preparation of disaster and reconstruction plans. There is a special need for hazard micro-zoning, along with preparation of hazard maps on large scale depicting risk variations. This would help in delineating high risk areas and thereby allotting them with more attention and precautionary measures.

Losses can be reduced by implementing improvements in design, maintenance and use of urban infrastructure, including its buildings and technology. Predicting and Preventing Disaster is essentially very complex job. Several techniques have been put to use in order to predict hazards. Important ones are *disaster simulation models, Delphi techniques and game simulations.* However, modeling of disaster should influence the communities and it may lead to changes in policy and a greater concern about safety through design. Disaster warning system has been a necessary component of any safety plan at the community level. Warning must be issued promptly and must be totally reliable. A valid warning system must consist of two discrete components, one highlighting the existence of danger and other focusing upon the ways and means to avoid or minimize the risks if not prevent it. Satellite communications links are generally are the best forms of prediction and subsequent warning. Proper warning systems allows for prompt monitoring of potential disaster agents and to permit evacuation or other preventive measures. However, to make warning systems effective, enough public awareness within the community is preferred.

**Development of a Community Safety Plan**

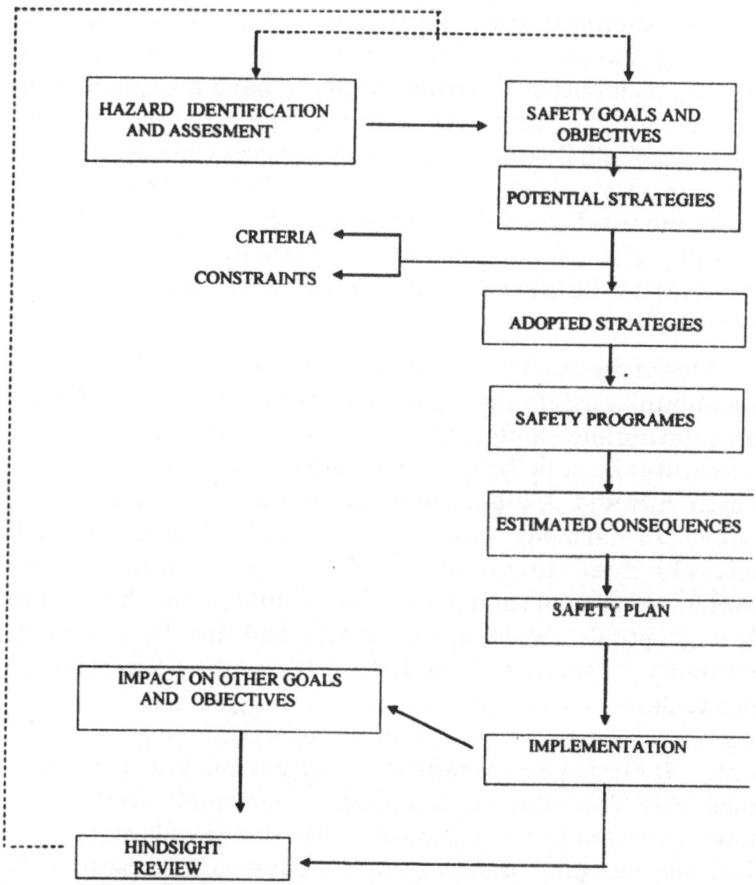


*Source:* Foster Harold D. (1980); Disaster Planning- The Preservation of life and property

## Community Safety Plan

It is clearly understood that natural disasters can lead to enormous economic losses and even ruin the economic growth entirely. Therefore, the need of the hour is to have a flawless community safety plan for every nation. The four aspects discussed above should form the basis of a community safety plan. The plan should include advance prediction, monitoring and warning of oncoming disasters, prepare the communities to face the challenging task of preventing the disaster and also help in chalking out the responsibilities of the administrative units. A community safety plan involves all pre-disaster planning to reduce the loss as a result of natural disasters. The four aspects, as mentioned earlier, needs to have a sequential and logical mixing to bring out the most effective outcome. There is also a need to follow an integrated approach while formulating a community safety plan wherein inputs from certain developmental planning aspects such as proper land use planning, risk micro-zoning to demarcate disaster-prone areas are needed to be taken up. All developmental projects should be targeted towards disaster mitigation.

One of the major challenges that comes out while preparing a community safety plan is the role of community itself. While formulating such a plan, there is no other alternative that can substitute the stakeholders themselves. After all, they are the mostly affected. A community safety plan, therefore, must be framed in harmony with the local needs. The communities actively need to be involved in the formulation and implementation of such plans. This should mainly be achieved through public awareness programs and also by following a bottoms-up approach. In order to attract their attention and also to derive their interest in the workings of a safety plan, it is quite essential to convince them regarding the plan content, which includes preparedness, mitigation and preventive measures. Therefore, such a programme should start with an honest attempt to make people realize their needs and the role that they can play in managing the adversities created by the disasters. Various community-based organizations and Self-help Groups are created for this purpose to promote effective and active community participation which will ensure local ownership addresses local needs and promotes mutual help to prevent damages.

It is also important to institutionalize the entire process which will encourage community involvement and therefore the involvement of government institutions and non-governmental organizations becomes crucial. They will facilitate and guide the entire scheme of things to develop a workable plan. Besides, government intervention motivates people stronger. It is to be kept in mind that active involvement from all groups of society, be it the community itself, or regional and international organizations, national governments or private agencies, will ensure prevention of natural disaster before they strike. The preparation of plans will start from formation of groups at regional level. These groups, thereafter, will carry out meetings, formal group discussions to reach out certain conclusive landmarks regarding the time of occurrence of hazards, the area mostly and estimate about the nature of resources available and similarly actions will be taken up consequently. Training of volunteers regarding social, resource, vulnerable and hazard mapping, seasonal analysis of different disasters etc. are carried out alongside. The aim is basically to ensure community participation for decision making and also to make the plans and programs, a sustainable process.

Every disaster plan should also include a detailed description of its purpose and the situations wherein it would be ideal. The co-ordinating committee while formulating the safety plan must be well aware of the disaster probabilities and the time-line of its occurrence. Several predictive models can be of good help. Each safety plan should incorporate simulations of probable disasters, valid prediction of its type and extent. A well prepared plan also focuses on the existing organizational structures, supervision an technical skills and improvements, time and again, is most essential. An effective safety plan should include list of all relevant officials, and their contact numbers so as to avoid extra panic situations when disaster breaks off. Spreading of disaster related information to the general public should also be a responsibility that a community safety plan should effectively perform. Besides, several post-disaster facilities and services such as search and rescue, community order, medical facilities, restoration of community services etc. are also important components of the safety plan.

What is important is to formulate a community safety plan that would be a product of local action, initiatives, local resources that would not only reduce the losses that takes place due to natural disaster but also to predict and prevent such calamities from future. A community safety plan to be beneficial and effective to all, proper public participation and continuous government intervention is desirable.

## Conclusion

Long-term planning and preparedness for disaster mitigation forms a part of the process of developmental planning. Science an technology inputs constitute its basic thrust, manifested in development of forecasting and warning systems, and post disaster restoration process. The traditional approach adopted earlier in managing disasters has been primarily concerned with disaster relief. It is only in recent times, that the gradual shift towards disaster preparedness and prevention catches limelight. This shift in approach has brought a significant positive change. However, to make such measures sustainable, it becomes essential to incorporate community participation and involvement in various stages of disaster management. The community action based plans and training, which forms an integral part of such an approach, improves community's problem solving skills and transparency in activities and sharing of knowledge ensures public participation.

*What is accepted by the community* should be of more important than *what is necessary* because the latter may not match with the local needs and desires, and may provide a negative impact. Formulating community safety plans will make the disaster management programs more feasible as local inputs will enrich its chances to get active support from the community as a whole. However, what seem to be most important is that the disaster mitigation plans should be institutionalized and linked with development plans, effective communication and use of latest information technology, extensive public awareness and education campaigns, involvement of private agencies to achieve maximum desired results.

## REFERENCES

Foster, Harold D.(1980); Disaster Planning - *The Preservation of life and property*.

Pandey, Vishnu and Okazaki, Kenji, *Community-based Disaster Management: Empowering Communities to cope with Disaster Risks*, UNCRD, Japan.

Prakash, Indu (1994); *Disaster Management - Social Vulnerability to Natural Calamities and man-made disaster preparedness and response* (Indian scene), *Rashtra Prahari*.

Rakshit, D.K. (1987); "Some aspects of community prepared plan against cyclones"; *Proc. US-Asia Conf. on Engineering for mitigating natural hazard damage*, Bangkok.

Sinha, D.K. (Eds) (1992); *Natural Disaster Reduction for Nineties: Perspectives, Aspects and Strategies*.

Smith, Kein (1992); *Environmental hazards - Assessing risks and reducing disasters*.

*The Report of High Powered Committee on Disaster Management*, (October,2001) National Centre for Disaster Management, IIPA, Department of Agriculture and Co-operation, Government of India.

*Training Module of Community Based Disaster Management - Research Project*, Supported by World Bank,(October,2005 to March, 2006), Aceh, Indonesia.

*Community-based Drought Preparedness Plan*, Government of Orissa, supported by UNDP.

# 12

# Biodiversity Management: A Case Study of Gautam Buddha Nation Park

*K.P. Goswami*

## ABSTRACT

The article reveals the study of wildlife diversity in the Gautam Γuddha national park. The major aim of this paper is to focus on the protection, preservation, development of wild life and also to provide a safe breeding place for wild life. The methodology used in the study of region is intensive field investigation on one hand and the data collected from the forest division and the wild life department of Gaya division on the other. The data are shown by cartographic methods. Forests around Danua and Bhaluas in Gaya district are known for its rich wild life resource. There are 34 small parks fall in the Barachatti range and 4 in Gurpa range. The area falls into three divisions Gaya Division (134.38 sq. km), Koderma Division (98.54 sq. km) and Chatra North Division (20.58 sq. km). Special features of the area are the existence of tigers. The area is spread over the either side of the Grand Trunk Road, which has made it easily accessible to poachers in the past, but wild life can still be seen in the area. The state wise census carried out in May 1972 indicated presence of two tigers in the areas. The merit of the area in providing the typical habitat for the tiger with the supporting species of animals is easily recognizable. The growing anxiety to save the depleted population of tigers resulted in the attention being focused on this area also. The discussion is made on these topics - shooting regulations, water supply, population studies, habitat investigations, forage development and control of threats.

**Key words:** Biological diversity, *Earth Summit, Conservationist,* Endangered Species, Shooting.

## Introduction

The 1992 United Nations Earth Summit in Rio de Janeiro defined 'biodiversity' as "the variability among living organisms from all sources, including, terrestrial, marine, and other aquatic ecosystems, and the ecological complexes of which they are part: this includes diversity within species, between species and of ecosystems". For biologists, biodiversity is the gamut of organisms and species and their interactions. Organisms appear and become extinct; sites are colonized and some species develop social organizations to improve their varied strategies of reproduction. For ecologists, biodiversity is also the diversity of durable interactions among species. It not only applies to species, but also to their immediate environment (biotope) and their larger ecoregion. In each ecosystem, living organisms are part of a whole, interacting with not only other organisms, but also with the air, water, and soil that surround them.

Biodiversity is not distributed evenly on Earth. It is consistently richer in the tropics (Kevin, 2004). As one approaches polar regions one finds fewer species. Flora and fauna vary depending on climate, altitude, soils and the presence of other species. For a listing of distinct ecoregions. In the year 2006 large numbers of the Earth's species are formally classified as rare or endangered or threatened species; moreover, most scientists estimate that there are millions more species actually endangered which simply have not been formally recognized. For example, Brazil's Atlantic Forest contains roughly 20,000 plant species, 1350 vertebrates, and millions of insects, about half of which occur nowhere else in the world. The Madagascar dry deciduous forests and lowland rainforests possess a very high ratio of species endemism and biodiversity, arising from the fact that this island separated from mainland Africa 65 million years ago.

The extinction of plants and animals is a biological reality (Ehlrich and Ehlrich, 1981). Today, all over the world, there is a growing concern for the protection of wild life which is fast disappearing due to man's trium-phant technological civilization resulting in large-scale clearance of forests to provide land for crops and for settlements, industries and transport networks. There has been a progressive deterioration of our planetary environment, which now demands a most careful look at the way man uses his habitat and adopts conservation policies (Dasmann, 1981). Wildlife constitutes all those animals which live in a *'wildscape'* - an area dominated by natural and semi-natural vegeta-tion, rock outcrops, rivers, glaciers, swamps and marine waves. These areas of wilderness are the natural habitats of wild life and they need a careful management to maintain small communities of plants and animals which establish their own ecosystem in them. Besides aesthetic, spiritual and ethical considerations to protect the *wildscapes*, it is now being recognized that they are essential to maintain the stability of the biosphere (Ehrenfeld, 1972).

The wildlife resources provide aesthetic, recreational and economic benefits and at the same time are important ecologically. Nature's millions of forms of life on this earth perform some or the other definite biological function which keeps the delicate balance of our atmosphere. In nature flora and necessary for the cross pollination of flowers and these with other animals also serve for dispersal and propagation of vegetation (Trivedi *et. al.*, 1994). Not only this, the health of vegetation also depends on natural browsing and grazing. Animals directly or indirectly depend on green plants for their food. Besides food plants also provide shelter to a variety of animal species equally important are the relations among animal population.

India has a rich heritage of wildlife. The region forms the greater part of the Oriental region of animals (Saharia, 1982). It has 350 species of mammals and 1200 species of birds of its own which are nowhere else found in the world. To conserve wild animals the various States of India have confined wildlife into national parks and sanctuaries (Nelson, 1971). The state-wise break up shows that nearly all the states have demarcated some areas for the preservation of wildlife.

The conservation is the practice embracing preservation, maintenance, sustainable utilization, restoration and enhancement of the natural environment (Holt, 1978). Wildlife conservation in this sense is the management of nature's living resources if we consider wild life as consisting both of wild flora and fauna. It also overlaps 'forestry' in so far as it is based on multiple lands use concept and implies management of wild plants as well as wild animals. However, under the present dispensation wild life management implies management of wild animals as distinct from forestry, and it implies maintenance of habitats (Singh, 1991).

The district lies in the deciduous forest zone where forest occupies 762.31 sq. km (1995-96). It is almost 15.65% of the total geographical area of the district. It is much below than the state level (21.4%) and far below than the standard level (33%) set under the National Forest Policy (Goswami, 2000). Since Gaya is enclosed with the hilly terrains and the sandy river Phalgu so it gets much heated during summer (Sinha, 2000). The main aim of the paper is to highlight the environmental issues of the study area and also mention the distribution of flora and fauna and their distribution, degradation and conservation.

## The Study Region

The study area falls geographically in the northern part of the Chhotanagpur plateau at Barachatti block in the Gaya district. The latitudinal extension of the district is 24° 17' 3" N to $25^0$ 3' 30" N and longitudinal extension is 84° 17' 35" E to 85° 23' 30" E (Fig. 11.1). The geographical area of the study region is about 253.50 sq. km. The region is surrounded by Mohanpur block (north), Hazaribag district (south), Nawada district (east) and Sherghati block (west) (Fig. 11.2).

There are two ranges (Barachatti and Gurpa) in which the park is extended. The region is formed by projecting high lands. Outcrops of highlands appear as hills, ridges and hillocks. The southern part is hilly region with a wide belt of grass wood jungle. Much of this part is barren and uncultivable in case of upward slope towards south. Climate is the most fundamental

**Fig. 11.1. Gaya District**

aspect influencing the economic resources pattern of any region that indirectly affects the settlement plan, agricultural works and other geographical activities. The study area is situated just north to the tropic of cancer (23.5° N). Gaya district enjoys monsoon climate for all practiced purposes, since the region is far from the ocean (424 km). The climatic data have been collected from the meteorological station of Gaya district and other block offices.

Soils are next to water, man's most vital natural resource. The pedology directly and indirectly bears the significant imprints of its vegetal cover. Vegetation protects the erosion of the surface. Roots of the vegetation act as binder of soil. The canopy of vegetation also attracts clouds for precipitation. It is the major source of building materials in rural areas. The distribution table of the given year is showing the variation of forest cover.

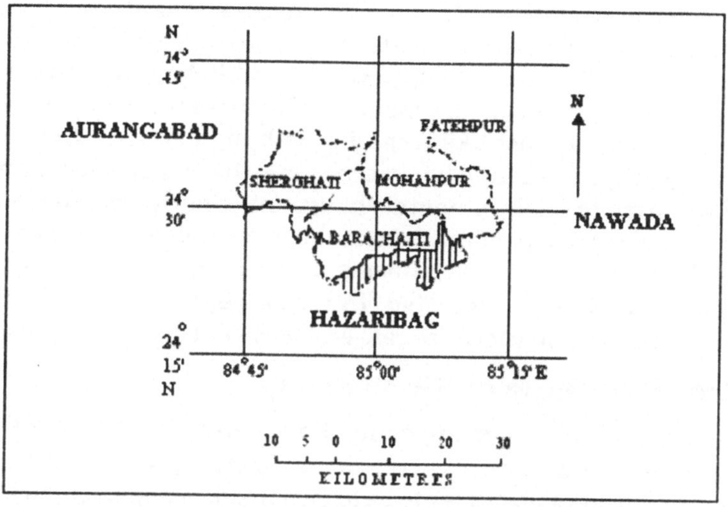


**Fig. 11.2. Location Map of National Park**

## Benefits of Biodiversity

Biodiversity has contributed in many ways to the development of Human culture, and, in turn, human communities have played a major role in shaping the diversity of nature at the genetic, species, and ecological levels. Biodiversity is what underlies many important ecological goods and services that provide benefits to humans. There are three main reasons commonly cited in the literature for the benefits of biodiversity.

## Ecological role of Biodiversity

All species provide at least one function in an ecosystem. Each function is an integral part of regulating the species balance, species diversity and species health: all aspects which are intrinsic for the ecosystem as a whole to survive and prosper. Ecosystems also provide various *infrastructure of production* (soil fertility, pollinators of plants, predators, decomposition of wastes...) *and services* such as purification of the air and water, stabilisation and moderation of the climate, decrease of flooding, drought, and other environmental disasters.

Research suggests that a more diverse ecosystem is better able to withstand environmental stress and consequently is more productive. The loss of a species is thus likely to decrease the ability of the system to maintain itself or to recover from damage or disturbance. Just like a species with high genetic diversity, an ecosystem with high biodiversity may have a greater chance of adapting to environmental change. In other words, the more species comprising an ecosystem, the more stable the ecosystem is likely to be. The mechanisms underlying these effects are complex and hotly contested. In recent years, however, it has become clear that there are real ecological effects of biodiversity.

## Economic Role of Biodiversity

For all humans, biodiversity is a *resource* for daily life. One element of biodiversity is crop diversity. Many see biodiversity as a reservoir of resources to be drawn upon for the manufacture of food, pharmaceutical, and cosmetic products. This concept of biological resources management probably explains most fears of resources disappearance related to the erosion of the biodiversity. However, it is also the origin of new conflicts dealing with rules of division and appropriation of natural resources.

Ecologists and environmentalists were the first to insist on the economic aspect of biological diversity protection. Thus, E. O. Wilson wrote in 1992 that: *biodiversity is the one of the greater wealths of the planet, and nevertheless less recognized as such.* Estimation of the value of biodiversity is a necessary precondition to any discussion on the distribution of biodiversity richness. This value can be divided into use value (direct such as tourism or indirect such as pollination) and non-use or intrinsic value. The concept of ecosystem services attempts to quantify the economic value to mankind of all the functions the natural environment performs.

Since biological resources represent an ecological interest for the community, their economic value is also increasing. New products are developed because of biotechnologies, and new markets created. For society, biodiversity also is a field of activity and profit. It requires a proper management setup to determine how these resources are to be used. Some of the important economic commodities that biodiversity supplies to humankind are: unique scientific research tools, food, medicine, industry, recreation and Ecotourism.

## Scientific Role of Biodiversity

This is a fourth benefit separate from the three above. Biodiversity is important because each species can give scientists some clue as to how life evolved and will continue to evolve on Earth. In addition, biodiversity helps scientists understand how life functions and the role of each species in sustaining ecosystems. The availability of unique genetic material for each living species may have incalculable value as evidenced by medical and genetic research that can lead to discoveries that may reduce mortality.

As of 2005 there have been numerous cases where genetic material unique to a given species has been utilized in developing a disease cure or producing a biochemical that is instrumental in medical research beneficial to humans. If genetic materials are lost through the present Holocene extinction event numerous medical cures will be foreclosed and lost forever.

## Discussion and Result

The forests around Danua and Bhaluas in Gaya district have been known to be rich in wild life. Special features of the area have been the existence of tigers. The area lies at the either side of the Grand Trunk Road which has made it easily accessible to poachers in the past but wild life can still be seen in the area. The state wise census carried out in May 1972 indicated presence of two tigers in the areas.

The merit of the area in providing the typical habitat for the tiger with the supporting species of animals is easily recognizable. The growing anxiety to save the depleted population of tigers resulted in the attention being focused on this area also. It was felt that this block of forest should be developed as a sanctuary with special stress on the protection of the tiger. Accordingly it was decided that a sanctuary named "Gautam Buddha Wildlife Sanctuary" be constituted by including a sufficiently wide and compact forest belt in Gaya, Kodarma and Chatra North Forest Divisions. The sanctuary was notified under the Wild Life (Protection) Act, 1972 in September, 1976. The divisionwise area is as follows (Fig. 11.3):

**AREA UNDER THE NATIONAL PARK**

| | | |
|---|---|---|
| 1. | Gaya Division | 134.38 Sq Km |
| 2. | Koderma Division | 98.54 Sq Km |
| 3. | Chatra North Division | 20.58 Sq Km |
| | Total area | 253.50 Sq. Km |

Source: Forest Division, Gaya

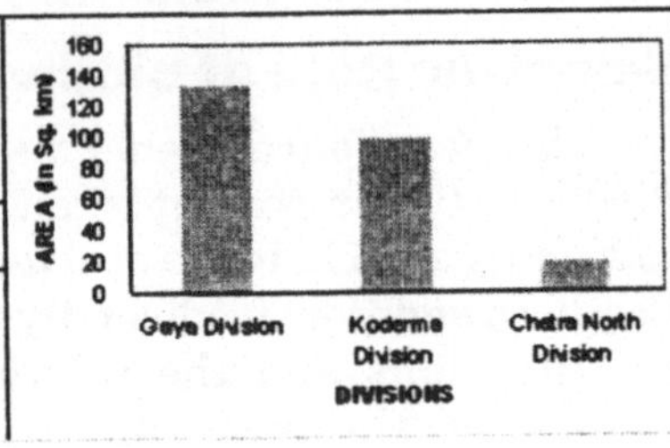


**Fig. 11.3.**

The following forests of Gaya Division forming the habitat of tigers have been included in the Gautam Buddha National park (Table 11.1).

**Table 11.1. Area under the National Park**

| *Range* | *Name of park* | *Thana & Thana No.* | *Demarcated Forest Area (in Hectare)* | *Demarcated Area Included Sanctuary (in Hectare)* |
|---|---|---|---|---|
| | Sabalpur | Barachatti 324 | 548.47 | 548.47 |
| Barachatti | Bhalua chatti | Barachatti 329 | 482.80 | 482.80 |
| | Burha Chaithia | Barachatti 323 | 1096.22 | 1096.22 |
| | Bhamdani | Barachatti 323 | 272.06 | 272.06 |
| | Tetaria | Barachatti 321 | 92.51 | 92.51 |
| | Sankhwa | Barachatti 322 | 868.54 | 868.54 |
| | Nagarchalia | Barachatti 314 | 572.99 | 572.99 |
| | Singhpur | Barachatti 319 | 564.92 | 564.92 |
| | Chootki Chapi | Barachatti 320 | 1177.18 | 1177.18 |
| | Barki Chapi | Barachatti 313 | 534.28 | 534.28 |
| | Sisiatorisalot | Barachatti 312 | 699.97 | 699.97 |
| | Harnahi | Barachatti 311 | 1026.07 | 1026.07 |
| | Bithiahi | Barachatti 310 | 98.86 | 98.86 |
| | - | Barachatti 316 | 326.17 | 326.17 |
| | Dang | Barachatti 315 | 198.88 | 198.88 |
| | Khaira | Barachatti 316 | 605.40 | 605.40 |
| | Phungunia | Barachatti 317 | 374.00 | 374.00 |
| | Piprahi Doat | Barachatti 309 | 794.80 | 794.80 |
| | Baijnathpur | Barachatti 328 | 2085.93 | 2085.93 |
| | Masaundha | Barachatti 333 | 268.59 | 268.59 |
| | Rangwa | Barachatti 652 | 3685.33 | 1842.67 |
| | Matgarha | Barachatti 654 | 2371.15 | 2371.15 |
| | Chitgarha | Barachatti 650 | 2248.93 | 1124.46 |
| | Keola | Barachatti 633 | 97.01 | 97.01 |
| | Kadarchuan | Barachatti 646 | 4638.25 | 2319.12 |
| | Malotari | Barachatti 642 | 550.30 | 550.30 |
| | Taliatanr | Barachatti 243 | 121.04 | 121.04 |
| | Hissir | Barachatti 258 | 1871 .58 | 1871 .58 |
| | Barsudi Guhi | Barachatti 259 | 677.76 | 677.76 |
| | Jaigir | Barachatti 260 | 1526.59 | 1526.59 |
| | Nare | Barachatti 283 | 1292.48 | 1292.48 |
| | Debobasara | Barachatti 223 | 2626.04 | 2626.04 |
| | Masrakh | Barachatti 284 | 754.63 | 754.63 |
| | Sarne Chorni | Barachatti 655 | 1504.51 | 1504.51 |
| | - | Barachatti 656 | 1727.47 | 1727.47 |
| Gurpa | - | Barachatti 557 | 481.40 | 481.40 |
| | - | Barachatti 658 | 236.54 | 236.54 |

Source: Forest Division, Gaya

## The Wildlife Management of Gaya Circle

**1. General Constitution:** All the forests of Gaya district have been included in this working circle. The term 'Wildlife' will denote the wild animals, mentioned in scheduled I to VII of the Wildlife Protection Act, 1972.

**2. Special Objects of Managements**: The special objects of managements are:

(*a*) To protect, preserve and develop wildlife:

(*b*) To provide for a safe breeding place for wildlife.

(*c*) To improve the carrying capacity of wildlife and

(*d*) To consistent with the above to harvest wildlife crop on a sustained yield basis.

**3. Management Units**: For management purposes, the forests of the district have been divided into five shooting blocks and a sanctuary in Barachatti Range. The sanctuary named as "The Gautam Buddha Wildlife Sanctuary" has already been notified under action 18 (i) of the Wildlife (Protection) Act, 1972 vide Government notification no. S.O. 1485 dated 14th September, 1976 for administrative convenience the shooting blocks have been constituted range wise. This will also provide sufficiently large areas to avoid intensive hunting.

## Shooting Regulations

(*a*) The shooting regulations prescribed in the Wildlife Protection Act, 1972 and the rules, there under will be strictly followed.

(*b*) Shooting in blocks which are under stocked with wild life may be restricted.

(*c*) Tiger and leopard feed upon deer, antelope, four horned antelope and pigs.

To preserve the tiger and the leopard it is necessary to keep a good stock of the animals they feed upon. While issuing shooting permits for herbivorous it will be necessary to consider the requirements of these carnivorous. Roughly there should be 70 animals available to the tiger for feed per year and half of this number may suffice for a leopard.

## Water Supply

The water supply during the rains and early winter remains adequate. The streams start drying up during winter and remain dry throughout the summer. However, Mohane maintains a scanty flow even during summer, all along its length through the sanctuary. A number of pools locally called *'dah'* formed in Mohane bed are almost perennial elsewhere throughout the area. Only a few water holes are the only sources of water supply during the summer.

## Wildlife Investigations

(*a*) **Population Studies**: The techniques of determining population indices followed in the Palamu National Park may be adopted with suitable modifications. The enumerators will be required to note flushing distances of animals, for calculation of tile effective width of the strip. Regular recording of tiger and leopards will be carried out in the area. To note how changes in the habitat influences the population structure and composition, observations will be made at water-holes or grass lands to keep a record of fawns (baby deer) soon after the birth period, at the beginning and the end of dry season and shortly before the next period of births. These observations will furnish data on fawn mortality or evidences of increase or decrease in the birth rates.

(*b*) **Feeding Habits**: The feeding habits of herbivores especially *sambhar* and *cheetal* will be studied. The wildlife guards on patrol will note the species and relative amount of vegetation, eaten by the wild herbivores every morning and evening. To provide an assessment of preferences, observation will be made to note the period or time for which a particular plant species is browsed or the frequency with which individual plants are taken.

## Habitat Investigations

The investigations will be needed to furnish data or production, availability and consumption of vegetative material within the habitat. Such study will be taken up in areas of special interest e.g. around water-holes, in areas of heavy use by domestic cattle.

(*a*) **Habitat Improvement**: For increase in prey, population, the habitat has to be manipulated for increased productivity. Dense forest cover provides little forage and poor covers to the herbivores. Hence, the forest canopy has to be opened where it is dense. Very large openings have, however, to be avoided as the herbivores do not stray much beyond the cover and consequently central portions of large openings remain unutilized.

(*b*) **Forage Development**: Pasturage will be raised in open areas by soil working, manuring and seeding for nutritional and perennial species. Fruit trees and fodder trees and shrubs will be planted in blanks at a spacing of 6m x 6m so as to encourage grass also.

(*c*) **Domestic Stock Grazing**: The impact of the grazing cattle is already quite heavy and hence against the interest of wildlife. At present not feasible to eliminate it, but every opportunity must be taken to curtail (cut short) it as much as possible. As a first step towards curtailing the impact of grazing, no cattle from outside the area will be allowed to graze in the sanctuary. Attempt should be made to get all the cattle inside the sanctuary inoculated against diseases that could be transmitted to wild life.

(*d*) **Water Storage**: There is shortage of water during the summer except in parts of the river Mohane. Steps should be taken to develop the existing water points for better detention of water and to create fresh water sources. Small dams should be constructed to create new sources of water. Tanks set in the ground can be a cheaper and useful device to collect rain water for summer, water-holes would be the cheaper device, but would be limited to river beds and would require regular maintenance during the periods of maximum use. Efforts would be made to ward off cattle as much as possible from the water sources.

# BIODIVERSITY: CONSERVATION, PRESERVATION AND PROTECTION

## Conservation of Biological Diversity

The conservation of biological diversity has become a global concern. Although not everybody agrees on extent and significance of current extinction, most consider biodiversity essential. There are basically two main types of conservation options, *in situ* conservation and *ex situ* conservation. *In situ* is usually seen as the ultimate conservation strategy. However, its implementation is sometimes unfeasible. For example, destruction of rare or endangered species' habitats sometimes requires *ex situ* conservation efforts. Furthermore, ex-situ conservation can provide a backup solution to in-situ conservation projects. Some believe both types of conservation are required to ensure proper preservation. An example of an in-situ conservation effort is the setting-up of protection areas. An example of an ex-situ conservation effort, by contrast, would be planting germplasts in seedbanks. Such efforts allow the preservation of large populations of plants with minimal genetic erosion.

## Species Preservation

This involves the protection of a particular species of wild animals on regional or at times on a global scale. For example, the black-buck is a protected species in the Bishnoi community of Rajasthan and herds of these deer roam about in the villages of this caste without any fear. The tiger is the national animal of India and is protected all over the country. The endangered species of Siberian crane is protected all over the world to save it from extinction, as only a few individuals hardly exceeding 300 are left on the earth. In the north east, Kaziranga National Park of India, the Great Indian Rhinoceros is the protected species. The manage-ment of such reserves is usually limited because of institutional and financial constraints.

## Wildlife Protection Act

In India, several laws have been made and sanctuaries and national parks are established for the protection of the dwindling wild life. India was probably the first country, to enact a Wild-life Protection Act. The Wild Birds and Animals Protection Act was passed in 1887 and repealed in 1912. For game protection in the states, in 1927, the 'Forest Act' XVI was enacted. Indian Board for Wildlife was established in 1952 and this was followed by setting up of Wildlife Boards in different states in India. In 1972, a new Wildlife Protection Act was passed. Under this act, possession, trapping, shooting of wild animals alive or dead; Serving their meat in eating houses, their transport and export all are controlled and watched by special staff (Chief Wildlife Warden). This act prevents hunting of female and young ones. Under this Act the threatened species are absolutely protected and the rest afforded graded protection according to their state of population size.

In India nearly 67 National Parks, 394 Wildlife Sanctuaries and 35 zoological gardens have been established for wild life management. National Park is an area which is strictly reserved for the betterment of the wildlife. Here activities, such as forestry, grazing or cultivation are not permitted. The total area of national park is 33988.14 sq. km. which is one per cent of the total geographical area. '*Sanctuaries are places where harvesting, killing and capturing of any animal is prohibited except under the orders of the authority concerned. It is strictly reserved for the protection of fauna.*'

Wildlife conservation includes all human efforts to preserve wild animals from extinction. It involves the protection and wise management of wild species and their environment. Some species have become extinct due to natural causes, but the greatest danger to wild life results from human activities. Thus we ourselves have created this need for wildlife conservation.

The progress of man throughout has been beneficial for the human race but it is the wild life that has suffered through the years. Hunting, Man's indifference and ignorance has led him to believe that he is independent of natural laws and of natural environments and forgets that the himself is a biological being and is directly and/or indirectly dependent on the natural resources available to him. He exploits, destroys and reduces the renewable natural resources, eradicates wild life and leads his own race to levels leading to catastrophe

In spite of all these management measures, extinction is the ultimate fate of all species. The purpose of wildlife management is to decrease the rate of extinction, which has unfortunately increased in the present century because of demographic pressure, indiscriminate hunting and poaching for valuable hides, skins and other materials.

## Conclusions

Human overpopulation and illiteracy are two of the major problems that bring about the tragic destruction of environment and animal extermination in the world. For every extinct or endangered species or sub species of animals, two of the major factors have been the destruction of habitats and excessive cropping by man. Several organizations are trying to bring about a successful turn-about regarding the concept of conservation. They educate and advise on the need for conservation of our depletion natural resources. All these organizations are run by men who are devoted and work their hearts out for saving habitats from destruction and species from extinction throughout the world. They fight against social, economic, political and religious problems in different parts of the world. All look like a hopeful sign that the future of the world's wild life is or growing concern to more and more human beings. But this is a race against time.

## REFERENCES

Dasmann, R.F. (1981). *Wildlife Biology;* John Wiley; New York.

Ehrenfeld, D.W. (1972), *Conserving Life on Earth;* Oxford University Press; New York.

Ehlrich, P.R. and Ehlrich. A. (1981), *Extinction: The Causes and Consequences of the Disappearance of Species;* Random House; New York.

Sagreiya, K.P. (2005), *Forests and Forestry*, National Book Trust, India, A-5 Green Park, New Delhi.

Goswami, K.P. (2000): An unpublished Ph. D. thesis submitted in MU BodhGaya entitled Impact of Ecology & Environment on Forests and Agriculture: A Case study of Gaya District, Gaya.

Holt, S.J.; and Talbot, L.M. (1978). *New Principles for The Conservation of wild Living Resources;* The Wildlife Society; Washington, D.C.

Jha, B.C. (1980): *Revised Working Circle Plan of Magadh Forest Division.*

Kevin, J. Gaston and John I. Spicer. 2004. *Biodiversity: an introduction*, Blackwell Publishing. 2nd Ed., London.

Nelson, J.G., Needham. R.D. and Mann, D.L. (eds) (1971), *International Experience with National Parks and Related Reserves;* Wiley; New York.

Saharia, V.B. (1982). *Wildlife in India;* Natraj Publishers; Dehradun.

Singh, Savindra (1991): *Environmental Geography*, Prayag Pustak Bhawan, Allahabad.

Sinha, Ashok Kumar (2000): A Geographical Analysis of Rural Markets and Rurban Centers, Published Ph.D. Thesis, M.U. Bodh Gaya. ESS. ESS. Publications, New Delhi.

Tuteja, T.K., Singh U.K., Sudershan, K'cherry and Trivedi, P.R. (1994): *International Encyclopedia of Ecology and Environment,* Vol. 4, New Delhi.

# 13

# Learning more about Private Teak (*Tectona grandis* l.f.) Plantations in Java: Promoting Better Income for the Poor

*Julia Maturana*

## ABSTRACT

Indonesian teak (*Tectona grandis l. f.*) wood production has been constantly decreasing along with the plantation productivity since the mid 1900s. A probable shortage of State teak wood could have a negative effect on the country and its people. Such effects could be buffered and/or eluded if the teak from private/community sources were to fill this gap.

This study included 54% of the total 82 districts of Java and used direct observations and surveys to show that private teak plantations are spread throughout Java. These plantations represent a very important source of teak although the current quality is considered to be lower than that from the State plantations.

Ensuring access to land and better prices for private/community wood through quality and market positioning strategies could support the development of new teak plantation areas and ensure an additional source of income for the millions of rural farmers in Java.

**Key words:** Private teak plantations; Java; income; production capacity; plantation system characteristics; teak farmer's profile.

## Introduction

The Indonesian island of Java is very well known for having the second largest planted teak (*Tectona grandis L. f.*) forest in the world, totaling more than 1.3 million ha in 2001 (Perhutani 2002b, Bhat and Ma 2004). This has helped Indonesia's teak production to play an important role in the international teak market, accounting for nearly 42% of the world's production in 2000 (Pandey and Brown 2000). In the national economy, it has also an impact in terms of goods production, earnings from exports and job opportunities. Internal sales for teak logs and sawn timber accounted for more than 680,000 $m^3$ and nearly US $87 million in 2001 (Perhutani, 2002b).

Globally, the teak supply is considered to be far below demand (Behaghel 1999, Maldonado and Louppe 2000). This demand gap and the high market price on the international market act as an incentive to increase the supply of teak wood. In Indonesia, this problem has pushed towards increments in illegally logged teak and reductions in the rotation period of the State plantations, both having an impact in today's production. Teak production from the State plantations has been constantly decreasing (Fig. 13.1) (Perhutani 2000, 2002a, 2002b) along with the plantation productivity since the mid 1900's (Wulfing 1932, Perhutani 1986) reaching one of the lowest production rates for these plantations.

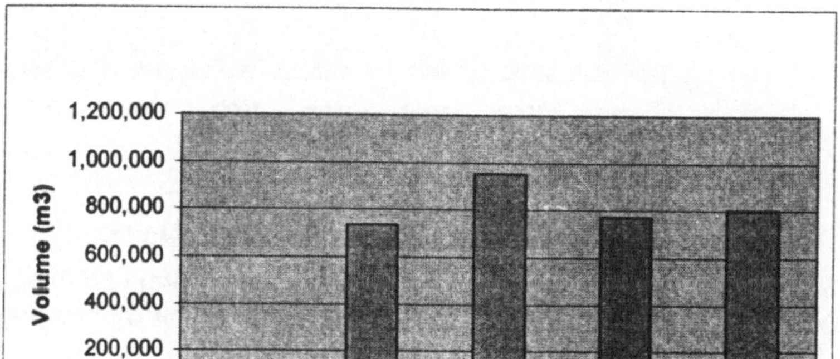


**Fig 13.1. Illegal logging from PP plantations in Java.**

*Source:* Perhutani (2002b)

Note: Information provided for number of stolen trees. Estimation based on an average volume of 0.3$m^3$/tree.

Illegally logging from Perum Perhutani[1] (PP) plantations has increased since the political changes in 1998, reaching a peak of 954 000 $m^3$ in 1999[2] (Fig. 13.2) (Perhutani 2002b). This number represents around 64% of the official volume legally produced by the State company PP in that year. The rotation period, on the other hand, has been reduced from the ideal of 80 years to 40 years (Perhutani 2003, pers. comm.; Maturana 2003, pers. obs.) to increase the number of plots that can be harvested and, therefore the volume of wood produced. State company officials have informally justified the shorter rotation period as a strategy for taking advantage of the plantations that would otherwise be illegally logged by the people.

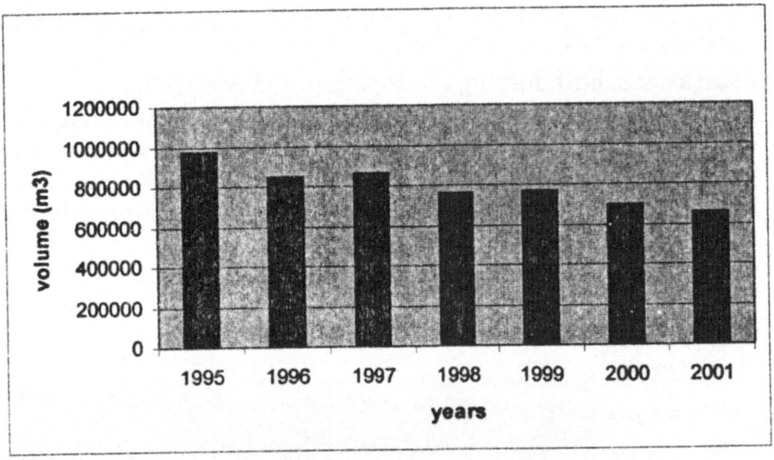


**Fig 13.2. Volume of teak Production in Java.**

*Source:* Perhutani (2000), Perhutani (2002a), Perhutani (2002b).
*Note:* The years 1995 and 1996 were estimated for Java using only Central Java data.

The present picture considers a potential shortage of teak wood coming from PP State plantations in the near future, with direct harmful effects on the captured earnings of the country and indirect effects through job reductions[3] in the downstream industry.

Private/community teak plantations are considered to be an important source of teak wood in the country; however the actual production capacity is not known and there is little information on the specific areas where these plantations can be found. Some studies have been carried out in specific locations like Wonogiri in Central Java and Kuningan in West Java as well as in the area of Gunung Kidul in Yogyakarta where the University of Gadjah Mada (UGM) has been conducting some research, but there is no information about the private/ community teak plantations for the rest of Java.

Participation of small holders and farmers in wood supply has become common in many countries (Bhat and Sharma 2003) in the world. In Indonesia, people have been planting teak for private use for a very long time and this tree plays an important role in the Javanese heritage (Peluso 1992). Nevertheless, it is only recently with new decentralization policies and regulations (Law 41/1999, Article 4; Law 22 on Regional Governance; Law 25 on Fiscal Balance) and the Sustainable Forest Council (SFC) policy for community-based forest management (CBFM) that the people's involvement in forest management is being supported. These changes have settled the legal framework for the establishment of privately owned teak plantations and enhanced them; several programs have been initiated and endorsed by the Ministry of Forestry (MoF) or local governments to encourage the planting of teak through the free provision of seedlings (MoF 2001, Fauveaud and Laburthe 2002, Merdeka 2002).

In these times, when planting and selling teak wood seems to be not only possible but facilitated for the Indonesian people, the private teak plantations could become an important source of supply for the already established teak wood industry.

The objective of this study is to learn more about these private plantations, in order to understand the right incentives needed to promote them as an alternative source of raw material for the industry and to increase the related income provided for the Javanese tree farmers. Specifically, this study has established a profile of the people who are planting teak trees and analyzed the characteristics of the different plantation systems observed, the capital requirements, the role of the market and other elements influencing teak farmers' decisions.

## Study Area and Methods

The study was conducted in 44 of the total 82 districts on the island of Java, including 12 districts in East Java, 20 in Central Java (including the area of Yogyakarta) and 12 in West Java.

The duration of the field work was about six months, divided in two parts: The observatory/preliminary phase was conducted from April to June 2003 and the remaining three months, for the interview phase, from August to October 2003.

We made a land trip to observe the extent and type of private teak plantations along Java, making a first selection of districts using the plantation's map of PP. This allowed us to identify the areas where State-owned teak is planted in Java and therefore those areas where private teak plantations are feasible.

The MoF local offices (Planning Division) provided information concerning the areas of private/community teak in their districts and the volumes of teak wood produced. The total area planted under the government programmes supporting teak plantations was sometimes provided. Besides the information about teak planted areas and the type of incentive programs in each visited district, we also gathered information related to the legal requirements for the selling process and prices.

Based on this information, sub-districts and villages were selected for visiting. When meeting with MoF local officers was not possible or when no information was available, we used the small roads to increase the probability of observing teak plantations while moving to the next district. The teak plantations observed included only those accessible by road using a four-wheel drive vehicle or in a range of a maximum walking distance of ½ an hour from the roads. The only area requiring boat transportation and about an hour's walk, was the area of Jati Luhur in the district of Purwakarta, West Java.

A total of 44 districts were visited. In each district we recorded the total observations of private teak areas, the kind of system, estimated age and other observable information. Whenever a teak monoculture area considered as planted was observed, the location was recorded using a Garmin 12XL GPS device. The main considerations to decide whether an area was or not planted were:

- Evident similar age of the trees
- Regular spacing of the trees
- Clear boundaries observed (not required).

Among the total 144 GPS locations recorded as private plantations, 90 locations were selected, equally distributed in the three provinces in 34 districts. In the second phase of the study, we made a visit to each of the chosen locations to interview the owner or manager of the plantation. A questionnaire was prepared to gather information about the plantation system, owner and ownership, market and capital.

## RESULTS AND DISCUSSION

### Production Capacity

A total of 23 district offices[4] provided information concerning the area, the number of trees and/or the volume of production of teak for East, Central and West Java at the sub-district level. These records were not very useful to estimate the production capacity of teak wood in Java given the low number of districts for which the information is recorded (about 72% of the total districts in Java are missing) and the scattered of the data[5]; nevertheless they are important in establishing that teak wood has been produced and traded with legal permits since 2000 from at least eight districts in Java, supplying the national industry.

### Kinds of Systems Observed

A total of 3437 observations of teak were recorded along the 5294 km of the trip including 44 districts in Java. According to the observations, teak trees are managed either in association with other trees/crops or in monoculture; in plantations or regeneration systems; and in four principal types of locations: **backyards; kebun** or forest gardens; **borders** and **fences**. The **backyard** system refers to teak trees found next to houses with or without other tree species. The **kebun** system refers to areas where no house was observed and teak trees were planted or managed in the areas either alone or jointly with several other species in fallow looking areas. The **border** system refers to teak trees bordering other tree plantations or cultivations, including rice fields. The **fences** refer to teak trees fencing properties, including or not other tree species. Besides these systems, teak trees were also observed in graveyards but with very low frequency, representing less than 1% of the total observations.

In general, teak was found associated with other tree species or cultivations, nevertheless teak in monoculture was also observed, representing 26% of the total observations. When in mixed systems, teak was observed mostly with seasonal crops such as: cassava (40%) and banana (19%). Other joint cultivations observed but in very low proportions were with tea plantations; mahogany (*Swietenia macrophylla*); sengon (*Paraserianthes falcataria*); mixed vegetables; sugarcane; coconut palms; oranges; acacia (*Acacia sp.*) and corn.

Most of the teak observed was not considered to be planted but regenerating in fallow looking areas. Teak plantations with spacing and homogeneous age represented only about 15% of the total observations, West Java showing the lowest proportion (6% of the observations).

Relative to the locations, the most common observation corresponded to teak in **kebun** systems (44.5%) mainly mixed with other tree species (72% of that) and the most common cultivations bordered with teak corresponded to rice fields (48%). Table 13.1 synthesizes this information.

## Profile of the Teak Growers

Most of the interviewed tree farmers, owners or managers of the plantations (a total of 78), were Javanese people natives of the area. Only two people were immigrants from nearby areas: Tegal in Central Java and Ciamis in West Java. This fact may be correlated with the high percentage of teak farmers (56%) who inherited their land or part of it.

### *Age*

About 63% of the teak farmers interviewed were people over 50 years old, most of them (42%) were between 50 and 59 years

**Table 13.1. Number of observations of teak in Java Provinces**

| *Province* | *Association* | | *System* | | *Location** | | | |
|---|---|---|---|---|---|---|---|---|
| | *Mixed* | *Alone* | *Planted* | *Regeneration* | *BY* | *KB* | *BR* | *FC* |
| Central Java | 683 | 505 | 276 | 912 | 448 | 429 | 152 | 159 |
| **Total obs.** | **1,188** | | **1,188** | | **1,188** | | | |
| East Java | 431 | 232 | 144 | 519 | 257 | 248 | 75 | 83 |
| **Total obs.** | **663** | | **663** | | **663** | | | |
| West Java | 1,422 | 164 | 99 | 1,487 | 418 | 852 | 65 | 251 |
| **Total obs.** | **1,586** | | **1,586** | | **1,586** | | | |

*Locations: Back yard (BY); Kebun (KB); Borders (BR); Fences (FC).

old; farmers under 30 years old were not represented in the sample. This fact may be due to the scarcity of land, we were often told that the first constraint for people wanting to plant teak trees was the availability of land. Indeed, people need to have enough space to plant crops before thinking of planting long rotation trees, some interviewed farmers mentioned that they could only plant teak when they finally inherited land or saved enough money to buy a cluster.

### Decision, makers

The decision to plant teak was mainly taken by the present generation of farmers and most of them were not from families that had been planting for several generations. A total of 88% of the plantations in Central Java; 89% in East Java and 71% in West Java were initiated by the present generation of farmers. In the remaining cases the decision had been taken by the father or grandfather of the interviewed.

### Jobs

In general (73%), teak farmers were observed to have an alternative source of income and to have links with the government and education sectors, PP or MoA. A very high number (39%) were businessmen, owners of shops or government employees and over 20% worked as farmers employed by other people to work in their fields.

### Income Ranges

We asked the farmers what their family budget was per month; in their calculation they included current monetary and non monetary inputs and represented the total as a lump sum or price. This amount varied depending on: if they still had children at school; if they had to buy food or if they grew food crops for themselves; if they could afford to eat meat; if they lived in the city or in a small village, etc. The results range from US $12/month to US $583/month with an average of US $112/month and a mode value of US $62/month. The results are presented in Fig. 13.3.

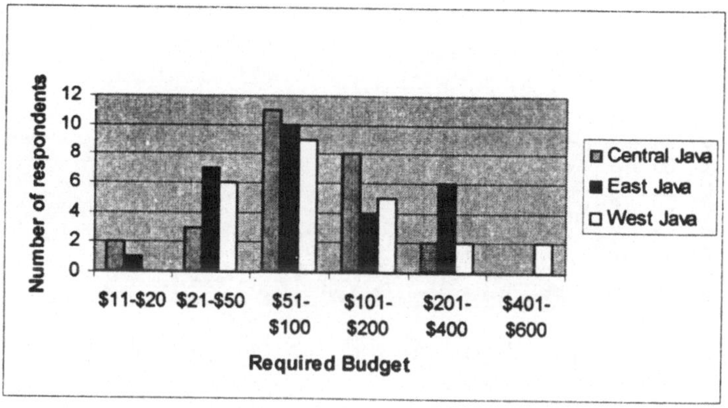


**Fig. 13.3. Teak farmers family budget per month.**

These results, confirmed with the perceptions from the field observations, show that most teak farmers have a relatively good financial situation and are not among the 37.1 million people (Relief Web 2002) considered to be living below the poverty line in Indonesia[6] (Relief Web 2002). In some cases the farmers told us it was difficult to live, but the living conditions of most were above average for their villages.

## Characteristics of the Plantation Systems

### *Ownership of the Land*

Most frequently (95% of the cases) teak plantations are developed on private land, although there were a couple of cases of plantations having been developed on public land such as alongside railway lines and along the roads. There were only two cases of teak planted on rented/loaned land from the State or the community, involving an annual rent of about US $2/year or tax payments of less than US $1/year.

### *Farm Sizes*

The average size of the areas was 2.17 ha, although the majority (49%) were under one ha and 40.3% under 0.5 ha in size. Few cases (six farmers or less than 10%) had areas larger than five ha with two farmers owning areas of 20 and 35 ha. The size of the areas owned is presented in Fig. 13.4.

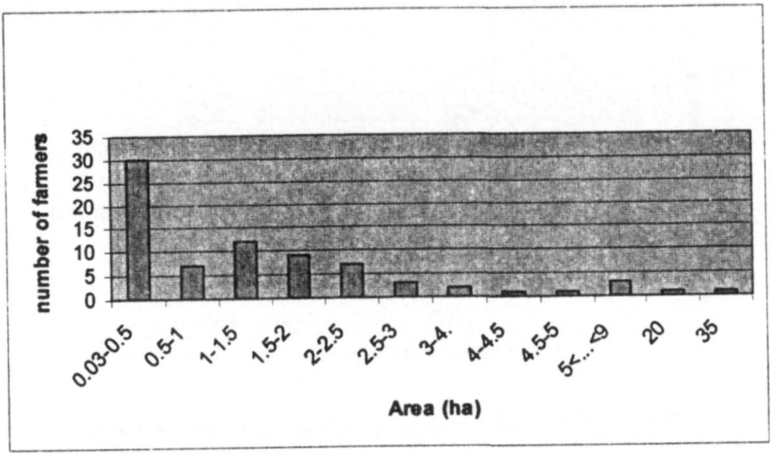


**Fig. 13.4. Size of the land owned by teak farmers in Java (ha)**

In Central Java, 85% of the teak farmers owned areas under 2.5 ha, 58% of which where under one ha; the remaining 15% owned between 2.5 and 20 ha. In East Java, 89% owned less than 2.5 ha and 11% had between 2.5 and 5 ha. In West Java, the owned areas were larger compared to Central and East Java, only 75% of the farmers had areas under 2.5 ha, with only 29% of those smaller than one ha; the remaining 25% represented areas between 2.5 and 35 ha.

These results are similar to those obtained by the Agricultural Census (BPS 1993) where 71% of the landowners owned less than one ha and 49 % of those corresponded to areas under 0.5 ha.

## *Plantation Area*

As would be expected given the size of the farms, the areas planted with teak were very small; the average size was 0.8 ha. The planted areas in West Java where the largest; this result may be correlated with the larger size of the farm areas in this province. As seen in Fig. 13.5, the area planted with teak, out of the total area is about 37%. Near 24% of the farmers planted all the area they own with teak: 31% of the farmers in Central Java and 25% in East Java and 17% of the farmers in West Java.

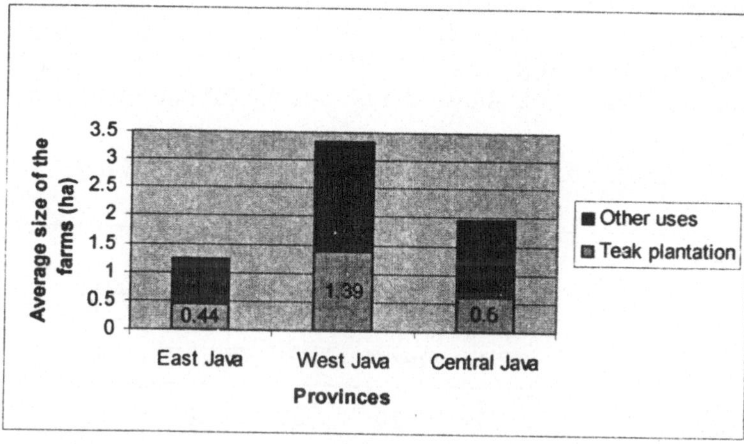


**Fig. 13.5. Average farm size and area dedicated to teak (prosince)**

## Kinds of teak

The tree growers mentioned several kinds of teak used to establish their plantations: local; superior or unggul; golden or emas; super; plus and bangkok, although the differences among them are not very clear. The main differentiation is the origin of the seedlings and rotation period.

The **local** teak is found in PP planted forests, with a suggested rotation period of 80 years. Farmers harvested the seeds directly from under the trees in the State plantations. The **superior** or **unggul** teak is the 'new' teak planted by PP in its plantations, it is a teak selected by PP for its ability to grow fast. This teak has an expected rotation period of 25 to 40 years and can be bought from PP nurseries.

Additionally, there are several cloned or 'fast growing' kinds of teak obtained through tissue culture: **golden** or **emas** teak, **super, plus** and **bangkok** teak. The origin of the seedlings is not completely clear but they most probably came from Thailand and Myanmar. These teaks are said to grow very fast; the different names depend on the different companies that introduced them to the market. The proposed rotation period for these trees is 15 years according to the companies' information; the seedlings can be bought directly from the companies or through distributors all around Indonesia at the relatively high cost of between US $1.2 to US $2.2/ seedling. Table 13.2 shows the details for PP and fast growing teak.

**Table 13.2. Production table for teak comparing PP and fast-growing teak**

| *Details* | *Company* | | | | | |
|---|---|---|---|---|---|---|
| | *Year 10th* | | *Year 15th* | | *Year 40th* | *Year 80th* |
| | *PP* | *Fast G.* | *PP* | *Fast G.* | *PP* | *PP* |
| Height (m) | 17 | 16 | 21 | 20 | 31 | 39 |
| Diameter (cm) | 15 | 25 | 20 | 35 | 38 | 58 |
| N/ha (trees) | 1,210 | 1,110 | 710 | 500 | 480 | 190 |
| Volume ($m^3$/tree) | 0.1 | 0.39 | 0.22 | 0.96 | 1.04 | 3.23 |
| Thinning/ha (tree) | 500 | 500 | 230 | 500 | 290 | 190 |
| Vol. thin. ($m^3$/ha) | 50 | 195 | 51 | 480 | 302 | 614 |
| Income (USD) | 3,000 | 28,537 | 3 050 | 175,610 | 85,000 | 172,220 |

*Source:* Perhutani 1986, Perhutani 2001, Monfori 2003.

Most of the farmers planted local teak (63%) but the 'fast growing' teak is present already in an important number of areas, mainly in East and Central Java (60% and 34% respectively).

## Capital Investment

A large proportion of the farmers (34%) did not invested capital to set up their teak plantations. These plantations were started with seedlings from the State plantations or other people's gardens and no fertilizer was provided or irrigation system required. The seedlings were planted just before the rainy season and the labor required to set these plantations up was the only input supplied.

For the farmers who invested capital, it was for the plantation establishment: buying seedlings, fertilizers and paying for labor workers. After the first two years the farmers would not invest further in any plantation maintenance, nevertheless in some cases, the farmers can afford to pay a guard to look after the plantation on a regular basis (as discussed below in the *labour input* section).

People investing in these plantations spent $784 US/ha in average, a large proportion of which was spent in seedlings and labor (37% and 22% respectively).

## Types of Systems and Density of Trees

Teak in monoculture is preferred in East and Central Java, where 68% and 50% of the farmers respectively planted teak in monoculture. In West Java farmers preferred to plant teak in a mixed tree system (79%); this system represented as well 19% of the cases in Central Java. Agro forestry systems were important in the three provinces, representing 31% of the cases in Central Java, 28% in East Java and 17% in West Java.

Farmers planting in monoculture explained this preference because of time constraints; they had chosen teak monoculture so they could be absent from the area and take care of other responsibilities. Monoculture systems showed densities of over 1700 trees/ha while agro forestry and mixed trees systems showed higher densities (> 2 000 trees/ha), with the exception of West Java, were the density was lower, because of the poor soil fertility.

## Former and Future Use

Most of the areas currently planted with teak are new teak plantations developed on land formerly dedicated to cash crops. These areas will probably remain as teak plantations in the future as the expressed intention of the farmers is to maintain these areas for the production of teak, even in the case of a crisis or sudden money shortages.

About 85% of the farmers expressed their intention to dedicate the areas to plant teak again or let it re-grow from the stump. In the case of financial need, almost half of them will maintain the teak plantations and look for additional sources of money; while 34% of the farmers would plant cash-crops and an additional 23% pretend to sell their land.

## Market and Capital Requirements

### *Source of Capital*

When capital investments were made (66% of the cases), almost all the farmers invested their own capital in the plantation with the exception of one case of an intra-family loan. No cases of bank loans or informal sources of capital were found.

### *Reasons for Planting Teak*

In the three provinces, the commercial aspect was mentioned as the main factor influencing the decision to plant teak; some people compared it to gold. Some others planted the trees for construction purposes (building houses) and their decisions were driven in terms of resource availability. For instance, many people mentioned that teak trees were very scarce in their region and they were concerned about the availability of building materials for their descendents.

Teak trees also seem to be the most popular tree to plant on non-productive land that does not receive enough water or is not fertile enough for food crops. The farmers believe that it is better to plant teak than having an empty space. In some cases, for the investors with additional sources of capital, teak was the best way to get wood and money without maintenance costs.

### *The Age of the Plantations and Expected Rotation Period*

Most of the teak plantations were planted after 1998 (73% of the plantations in Central Java; 64% in East Java and 25% in West Java), the youngest ranging from six months to nine month old, pointing to the possible influence of the political and legal changes with respect to teak and communities. The oldest teak plantation observed was 40 years old.

The expected rotation period was short, ranging from 20 to 35 years, which might also correlate farmers' decisions with the appearance of a bunch of private companies such as Monfori[7] which specialize in supplying 'fast growing teak' seedlings. These companies claim to provide teak seedlings that can be harvested after a 10-15 year period with average yields of 400 $m^3$/ha in the 15th year.[8]

### *Labor Inputs*

Although most of the teak farmers (94%) lived near the plantation area, almost half of them (44%) sub-contracted labor for the settlement of the teak plantations: 42% in Central Java; 54% in East Java and 33% in West Java, the rest of the cases corresponded to family labor. While this decision may have socio-cultural roots, we must also recall that a high percentage of these tree farmers have additional occupations and sources of income, which may explain the need and capacity for hiring labour.

### *Markets and Prices*

Most farmers who want to sell their wood would wait for the buyers to come and make an offer for their standing trees. The reason given for this passive attitude was that they wanted to avoid getting involved in the paperwork (harvest and transportation permits) even though these permits were easy to obtain and the process was very quick and cheap.[9] The second reason was the well established market for teak throughout Java; traders often visit the areas offering to buy the wood sometimes even before the trees are ready to harvest.

The prices were referred to in terms of US/tree and not in terms of $m^3$ and usually for standing trees. Farmer's general understanding is that a 20 years old teak tree is equivalent to 1 $m^3$. The expected average price was $150 US/tree and the range goes from US $6/tree to US $583/tree and in general, people did not have a fair expectation of prices but expected instead to make *enough money*.

## CONCLUSIONS AND RECOMMENDATIONS

Private teak plantations can be found throughout Java in a wide range of ages, from six months old up to 40 years old; representing a large potential source of industrial raw material for the country. Private teak wood from at least eight districts in Java is already supplying the national industry and we observed teak plantations in all 44 districts visited, where teak farmers conveyed commercial reasons as the main factor influencing the planting.

Because of its wide distribution, private teak plays an important role in supporting many small farmers and poor people, through direct financial gains, income substitution and the creation of jobs. One teak tree sold at $150 US dollars would represent an income of over three times the country's average monthly wage[10] and over two times the expressed amount of money required for living. The benefits provided by the teak plantations, calculated as the present value of the expressed price per tree, tree density and area planted for a rotation period of 30 years is approximately $290 USD/month;[11] almost seven times the average wage for Indonesia.

Relative to its significance in providing jobs, approximately 44% of the teak owners subcontracted local people as laborers. Although the period was restricted to the first two years, it may represent an important job source for poor people, given the wide distribution of these plantations. Additionally, privately owned teak wood is the preferred source of raw material among the Javanese for building houses and producing local furniture. This generates further employment and substitutes an important portion of people's expenditure.

Land availability seems to be the main factor restraining the expansion of teak plantations; these plantations developed on private lands occupy a large percentage of the farm areas (about 37%), but this corresponds only to an average size of 0.8 ha. An increment in the farm areas could imply an important increment of land dedicated to teak in Java.

The lack of land is also restraining the development of teak plantations through its effects on price. Private teak is mainly found in mixed systems, planted among high densities of other trees or crops on small patches of land, back yard areas or bordering cultivations. These high densities, along with the lack of fertilizers and pruning and thinning practices result in thin, crooked trees with multi branches, which do not fit with the requirements for the furniture industry exports. The current price paid per $m^3$ for this wood coming from private sources is about half the price paid for the wood coming from the State plantations, mainly because of the perceived differences in quality.

The West Java region presented a special case. This area, when compared with the other two provinces, was characterized by larger farms and planted spaces. Local teak planted in low densities dominated in this region, with mixed tree systems preferred to monocultures. Most of the farmers worked with family labor and did not make capital investments to establish the plantations. Those farmers that did, however, were characterized by having made the lowest investments per ha. These results may be due to the fact that this province is characterized by lower soil fertility and dryer areas and does not allow for the establishment of cash crops; the main competing use for the land on the island.

Private teak plantations in Java will be an important source of teak wood for the country, as these plantations seem to be increasing. This probably is a result of the new policies and regulations promoting people's involvement in the management of natural resources, as well as the appearance of new seedling provider companies in the market. About 70% of the plantations in East and Central Java were established after 1998, most of them with new planting materials.

It is probable that wood from private plantations will substitute some of the State wood sources in the nation's teak processing industry. This is likely, given the decreasing volumes of teak wood coming from State plantations, the demand gap for this wood in the international market and the availability of the main inputs (teak seedlings and labour) required for the establishment of private teak plantations in Java. To guarantee that such substitution is beneficial for the farmers would require ensuring better prices and quality for the wood coming from private sources. Today, the quality of wood from private sources is not perceived as being appropriate for the production of export-oriented garden furniture; the most important contributor to PP earnings (Perhutani 2002b). This affects the price and the demand of private teak.

Supporting the plantation's management and maintenance promoting fertilizing, pruning and thinning techniques, would improve productivity rates and wood quality. These factors would in turn lead to better market prices. Another favorable strategy would be to develop 'farmer associations' to sell the wood directly to the processing companies, rather than through intermediaries, and to reduce the number of farmers involved in negotiating harvesting and transportation permits. Currently, most of the farmers sell their teak wood as standing trees in their areas, thereby loosing the price difference that could be obtained if they were to take their wood to market themselves. Finally, the use of communal and poor-quality lands could serve as a viable option to overcome the lack of land, which restrains the further development of teak plantations and its related benefits for the millions of rural farmers in Java.

## NOTES

1. A State Company that manages near three million hectare of forests and forest plantations in Java.
2. Estimated from the quoted number of trees logged and an average volume of 0.3m$^3$/tree (Perhutani 2002b).
3. A recent survey made in Central Java including 91.43% of the 35 districts and 8183 formal and informal furniture producing clusters, showed that these establishments had an average of 5 workers not including family labor (Sulandjari and Rupidara 2003).
4. A total number of 39 district offices were visited but not all of them had information referent teak in the community forestry areas.
5. Detailed information available from the authors.
6. Estimation based on an average of four people per home and poverty limit of USD 4.84/person/month (Bureau of International Labor Affairs 2003).
7. Monsanto created Monfori in 1999, to sell Mahogany seedlings, hybrid eucalyptus, and a selected teak clone called Jati super. This company was the first to introduce this type of teak. Monfori became independent in 2002.
8. Based on a density of 575 trees/ha after thinning in the 7th and 10th year (Monfori 2003, PPKKI 2003).
9. As mentioned by the MoF office of Boyolali, Central Java (Supriyadi 2003, pers. comm..); the permits cost only USD 0.12/

m3 each: the harvest permit or Ijin Tebang and the transportation permit or Surat Keterangan Saanya Hasil Hutan (SKSHH).

10. The average wage for Indonesia is $42 USD/month (AIG 2004).
11. Using an average price of $150 USD/tree, assuming a density of 1 000 trees/ha, 0.8 ha planted, and deducting the average initial investment of $784 USD/ha in the first two years for a time period of 30 years. The discount rate used corresponds to that which applies for savings at commercial banks (12%).

## REFERENCES

AIG, 2004. *Regional Risk Analysis for Asia Pacific*, Indonesia.

Behaghel, I. 1999. Etats des plantations de teck dans le monde. *Bois et Forets des Tropiques* 262(4): 6-18.

Bhat, K.M. and Sharma. J.K. 2003. *International Conference on Quality Timber Products of teak from Sustainable Forest Management*, KFRI, Kerala, India.

Bhat, K.M. and Ma H. O. 2004. *Teak Growers Unite!*. *ITTO Tropical Forest Update* 14(1): 3-5.

BILA, 2003. The National Poverty Line in Selected countries, official thresholds and other estimates, most recent year. U.S. Department of Labor - Bureau of International Labor Affairs.

BPS, 1993. *Sensus pertanian* 1993. Jakarta, Biro Pusat Statistik.

Fauveaud, S. and Laburthe. P. 2002. Etude de la filiere teck a Java centre: Des industries du meubles de Jepara aux plantations de teck. Paris, *Institut National Agronomique Paris-Grignon*: 77.

Maldonado, G. and Louppe. D. 2000. Challenges of teak in Cote d'Ivoire. *Unasylva* 51(201): 36-44.

Merdeka, 2002. Perhutani Akan Reboisasi 25.000 Ha Hutan di Jateng. *Suara Merdeka Newspaper*.

MoF, 2001. Seed for People - Surat Menteri Kehutanan No. 973/Menhut-V/2001 tanggal 29 Juni 2001, Ministry of Forestry.

MONFORI, 2003. Spesifikasi Produk Jati Super Monfori. Jati Super, Monfori Nusantara.

Pandey, D. and Brown. C. 2000. Teak: a global overview. Unasylva 51(201): 3-13.

Peluso, N. 1992. Rich Forests, Poor People: resource control and resistance in Java. Oxford, University of California Press.

Perhutani, P. 1986. Tabel-Tabel bahan: Latihan kerja penataan hutan. Madiun, Biro perencanaan, Perum Perhutani Unit II Jawa Timur.

Perhutani, P. 2000. Buku saku statistik tahun 1995-1999. Semarang, Perum Perhutani Unit I Jawa Tengah.

Perhutani, P. 2001. Pedoman Pembagian Batang Kayu Bundar Jati Tahun 2001. PHT 91 Seri Produksi 136. Perum Perhutani. Jakarta: 9.

Perhutani, P. 2002a. Buku saku statistik tahun 1997-2001. Semarang, PT. Perhutani (Persero) Unit I Jawa Tengah.

Perhutani, P. 2002b. Statistik tahun 1997-2001. Jakarta, Direksi PT Perhutani.

PPKKI, 2003. Kultur jaringan. Jati. Jember, Pusat Penelitian Kopi dan Kakao Indonesia.

ReliefWeb, 2002. Indonesia Appeal No. 01.61/2003. R. Web. 2004.

Sulandjari, S. and Rupidara N. S. 2003. Analisis Rantai Nilai pada Klaster Furnitur di Jawa Tengah Value, International Labor Organization & Centre for Micro and Small Enterprise Dynamics (CEMSED): 42.

Wulfing, W. V. 1932. The sample-plot investigations of A.E.J. Bruinsma; yield tables for teak plantations. Tectona, Indonesian Forest Research Institute Special Publication No 30a.: 1454-1509

# 14

# From Social Capital to Profits: Discourses on Corporate Ethics, Governance, Choice and Values

*Dr. Prasenjit Maiti*

## ABSTRACT

The concept of social capital was originally conceived in the area of the social sciences, especially in Political Science and Sociology, to understand the networks of trust and cooperation that facilitate human interactions, transactions and exchange in the sphere of the civil society. This article argues that the theoretical category of Social Capital can also be suitably utilized in the domain of the behavioral sciences, especially in Management Science, to understand various angularities of disciplines such as Industrial Relations, Labour Welfare, Human Resource Development, Organizational Behaviour and Supply Chain Management. It is critical to appreciate that Social Capital is not only an altruistic concept but that it is also a tool of self-preservation and profiteering that happen to constitute the core predatory instincts of conscientious business managers and their industry-specific think tanks.

**Key Words:** Social Capital, Rational Choice, Value Management, Culture Root Paradigms, Sociability Discourse, Industrial Relations, Inclusive Growth, Sustainable Development, Corporate Social Responsibility, Corporate Governance, Multi-Stakeholder Dialogue.

## Introduction

Social Capital as an analytical tool can be used in the field of Business Management to facilitate different business practices from both an ethical angle and a utilitarian perspective (Bjørnskov 2004 and Beugelsdijk 2004). Business managers should try to entrench networks of trust, loyalty and cooperation within and without their organizations. This also makes good business sense in terms of rational choice theory.

Good business is also about establishing customer-friendly images in a manner that highlights values such as reliability, trustworthiness, quality, economy and durability. These values evolve over time and are underpinned by a sustained relationship of confidence. So specific products more often than not become identified with brand names either for niche or for broadbased markets. Brand equity and positioning so generally depend upon the successful merchandizing of products and their images (Boix and Posner 1998).

## Theoretical Framework

The *problematique* of this article is structured around the polemic of civil societal institutions (networks and embeddedness) and democratic governance (inclusive growth and participatory development) among other critical areas of social scientific research. This line of theoretical research is expected to contribute new knowledge and facilitate innovative research to better understand the interactions and interplay between actors and their institutions (Bourdieu 1979).

The politics of everyday life and human development are generally informed by the dynamics of choice and the strategies of cooperation. This tension can be somewhat resolved by adopting the social capital approach, as inclusive growth can more often than not be ensured by empowered choices and delegated actions.

The above contention is supported by the World Bank's perception of social capital. The Bank tends to apply the lessons learnt from various projects funded under its auspices in the light of community fabric, collective trust and networks of cooperation sustained by the ethos of multi-stakeholder dialogues.

This highlights the importance of social capital in all our grand as well as small narratives today in a world where economic signifiers and their signified meanings are more or less in a state of constant flux provoked by the marketplace of politics and the social as well as cultural rhetoric of glocalization.

The Bank has also championed the Think Globally Act Locally slogan while the Millennium Development Goals themselves are a construct of social capital that require meaningful negotiations with the perceptions of disengaged civil societies and panoptic state systems.

## Tradition of Trust

A perspective analysis of the dynamics of trust in any business organization happens to be a most problematic exercise that has to be carried out over a period of time and that involves multiple stakeholders as well. Why should one trust his / her peers, superiors and subordinates? Is trust a construct of culture specifics? Can trust be equated with the given socio-politico-economic realities of any spatio-temporal context? Can trust be learnt or emulated? Can trust be analyzed in terms of cost-benefit calculations?

These are rather disquieting questions with no unilinear answers. Business leaders have to realize that the pedagogics of trust and resilience of professional relationships can only be tested against either hypothetical or real life situations where the actual motivations, aspirations, perceptions, preferences and culture root paradigms of individuals or groups are explicitly exposed in the given context of their informed self-interest (Bourdieu 1986).

Trust happens to be a matter of choices that is not altogether different from the various other choices we make during the course of an ordinary day. But choices are also inspired and conditioned by values that are established standards of social interactions and constantly relearnt during the life of organizations and the professionals who sustain such organizations.

But trust is also a societal resource that is limited by its very nature, as individuals more often than not put their own self-interest first rather than the interest of their respective organizations. The core challenge is to translate these small narratives of power into a grand narrative of preference that would ultimately facilitate a transition of business culture of given organizations.

So the game that is to be played by the different parties concerned would be to transcend the *actors and institutions* model to the *actors in institutions* prototype. This would be a game that requires a multiplicity of stakeholders and a plethora of their interests that would finally be integrated in an overarching design of motivations, aspirations and role performance (Burt 2000).

When disjointed players start to commence upon the game upon a somewhat level play field then emerges the ever-critical issues of entitlements and capabilities. Each and every player has got entitlements to improve upon his / her own material conditions of life, status of empowerment, value systems and catalyze the politics of everyday life. These entitlements have then to be qualitatively upgraded to the level of capabilities. So from professionals to stakeholders is a long journey of accomplishments that would institutionalize a culture of trust and cooperation in organizations for the mutual benefit of all the concerned stakeholders.

## Trust and Role Playing

Trust and cooperation happen to be dynamic aspects of human behavior that are both acquired as well as achieved during the course of everyday interactive exercises. One has to appreciate that trust as an interface is perhaps more potentially real and tangible rather than trust as a construct, as the former is an arrangement of convenience, connivance and knowledge while the former is a living entity that comes across our daily social interactions as a concept that works in real time. In essence, therefore, reciprocation sustains trust while trust facilitates cooperation.

Why should one cooperate with another? Is it a learnt reflex or is it based on enlightened self-knowledge and self-interest? Would cooperation lead to intense creativity, new images, different ideas and better emotional and stress management? These questions and their answers have to be delved into and dwelled upon before we even try to garner an entire array of new knowledge on this subject.

Relationships can be built upon cooperation, competition and conflict. Each such mode will explore new models of interface and would finally lead to an inchoate world of new possibilities where men and women in the industry would steadily learn how best to unlearn past lessons and deschool themselves from the cultural baggage handed down the generations.

For one lesson is almost clear at this juncture of empirical findings i.e. actors will only cooperate if there are elements of confidence and potentialities of profit to be found in such bilateral or even multi-modal discourses. The architecture of social knowledge that fosters industrial relations and the microphysics of power that acts as a conducive ensemble in the background of labor welfare are, after all, situation-driven and culture-specific realities that need to be continuously re-examined against the context of changing politics and economics of the body social (Burt 2007).

But what about the body corporate as a whole? This is an entity that can neither be enticed nor be allured into rash decisions and myopic strategies. If cost-cutting is the order of the day and free rider expenditures are on their way out, then it is also the duty of business managers to adequately sensitize their industrial work forces and integrate them in a culture of consensus without which the entire complicated domain of trust management would be severely disempowered.

## Negotiating Trust

From the management of human values to the management of corporate trust is a long, arduous and unenviable journey that requires out of the box thinking matched with the capacity to relate to the empathy of individuals. So rapport-building would appear to be a crucial requirement in this context of trust management when business leaders should be able to overcome the tedium of Human Resource Development overkill in order to efficaciously enter into revolutionary dialogues with both blue as well as white collar workers to make them chant the mantra of their respective organizations underpinned by their rational choices (Chloupkova *et al.* 2003).

## Ideas/Images

It may appear to be rather far-fetched to deploy a highly specialized social scientific theoretical category like Social Capital in the area of Business Management and Industrial Relations. The latter is a network of relationships that enter into professional dynamics with one another in the highly charged world of business informed by factors such as the profit motive, ethical applications, Corporate Social Responsibility, conflict resolution and a sense of accountability *vis-à-vis* the environment, sustainable development and inclusive growth.

So innovative ideas supported by creative images are required to streamline Industrial Relations in the post-globalized scenario that predicates upon both a breakdown of barriers as well as establishment of new and difficult barriers in the cyber world of netizens, their perceived expectations and their ever-increasing sense of achievements and complex role performances (Coleman 1984).

So we have to understand the *problematique* of the increasing hiatus between business managers and their employees in the context of relative deprivation that may be either incremental or else decremental in nature. The critical ploy that has to be strategized in this connection is the network of inter-dependence that depends upon the fluctuating degrees of trust and cooperation and overall organizational health of any given industry.

## Rationality of Choice

It is somewhat difficult to estimate the rationality of a choice until the outcome of that choice is perceived either in material or else in psychological terms. Exactly *how rational a choice is* would have to be decided subsequently when the question *how utilitarian that choice was* would emerge in the future. The issue of Industrial Relations is all about the management of choices in the workplace. Why should workers obey the management? What are the concerns of the management that are reflected in its style of leadership? Is loyalty related to motivation? Or is loyalty a function of the chain of command? What are the effective outputs of hierarchy?

The entire discourse on *Discipline and Punish* is predicated upon the assumption that the State (in this case the Management) is always monitoring and evaluating the actions (and the consequences of such actions) of its Citizens (in this case the Workforce). This social scientific approach to conformity rests on the hypothesis that individuals as rational actors would intelligently choose between the options of conformity and non-conformity. While the former leads to the area of Reward Management the latter leads to theories of penology i.e. how to mete out punishment that would rectify and improve upon i.e. re-condition errant behaviour.

So the question of effective communications arises at this juncture. How effectively can managers communicate with their workers is a serious issue. Here emerge the concerns of dialogue versus pedagogue. What are the qualities of dialogue that distinguish the channels of communications from the aspects of the power-authority discourse?

## Choice as a Paradigm

The model of multiple preferences is rather complicated, as it entails different types of choices in different space, time, culture, knowledge and power discourses. The choices before a Trade Union may not always be uniform, and may be influenced by considerations like a global economic meltdown, local political tension, national emergency, social problems, regional discrepancies in development, bad economy of scale, retrenchment, cost-cutting etc.

But the critical central issue still remains the anatomy of choice: why do men and women ordinarily behave in one manner and why do they tend to behave otherwise under duress? This in itself constitutes an interesting universe of research where various factors such as values, beliefs, attitudes, inclinations, orientations, opinions etc come into play in the wider area of choice influenced by an inchoate regime of deterministic / indeterministic chaos (Coleman 1988a).

## Ownership/Trusteeship

The management and the workers have to mutually entrench an ethos of cooperation that would both create and sustain the values of Ownership / Trusteeship in their given organizations. One has to belong to and long for one's organization simply because organizations are the institutions of collective behavior where people associate and interact to pursue certain set goals.

The very fact that these targets have to be achieved and new targets innovated point to the fact that the organizational health of any given industry is sound. The point of saturation has to be ably and creatively transformed into an arena of new possibilities that lead to new challenges and innovative gestures.

But achievement of targets is critically dependent on team performance, and so the team players have to share different degrees of responsibilities and have to be burdened with different measures of failures and successes. The most important consideration at this point is to determine who can shoulder what amount of burden. So the question of authoritative allocation of values has also to be pondered upon actively by the management concerned.

The issue of Ownership/Trusteeship is also almost organically related to the issue of institutions. Each business organization, in sociological terms, is an institution replete with a history of evolution, work culture, Trade Union ethics and typical management practices. So the study of business organizations as dynamic institutions of power, authority, repression, obedience, reward and punishment may lead to an ulterior understanding of Industrial Relations.

Actors and their interventions become criticalities that have to be objectively factored into the discourse of polemics that center on issues/concerns of Industrial Relations. The credo of loyalty is also another ponderable that may be studied or else addressed in order to appreciate the different dynamics of Social Capital, Rational Choice and Value Management.

## Value Management

The question of moral values and business values may occasionally come across as a dichotomy, especially when the role of social values such as commitment, traditions and loyalty is concerned. The recent economic recession worldwide has prompted Big Business in India to retrench, cut costs, economize on operations, production, marketing and other strategic areas of their daily activities. But there have been instances where probationers given the pink slip were again brought back home, so to say.

This is perhaps a truly Indian phenomenon that has been also somewhat motivated by external political pressures. But this event is not to be identified with the somewhat out-of-vogue concept of Asian Values. The issue of Industrial Relations as sustained (hypothetically) by the arrangement of trust and cooperation in business is also perhaps a function of traditional Indian family values that happen to rely much more on loyalty and feudal ascriptions rather than the rational-legal structure of legitimacy and authority (Coleman 1988b).

## Change as Resource

Organizational change can be perceived as an industrial relational resource that can be utilized by different stakeholders to add value to their concerned industries. It is not always quite easy to address change in organizations, as multiple players find it more often than not difficult to cope with transformations and transitions. Organizational development, evolution and health depend upon targeting moving goals that are continuously in a state of flux.

But how really to use change as an organizational resource? Changes can take place in terms of qualities of output, levels of production, modernizing technologies, innovative methodologies, deschooled pedagogics and radically different marketing strategies. Such changes vis-à-vis organizations can even be studied in the context of the Guest-Host discourse where changes are not perceived as ghosts of spatio-temporal realities but rather as guests in the context of host organizations.

So change can also be sustained as a resource in the context of organizational change and metamorphosis. Whenever changes occur, there are sweeping as well as drastic effects on the levels of motivation, business strategies, styles of leadership performance, marketing techniques, economies of scale, interpersonal relationships, priority areas and preferential values. It is imperative that this phenomenon of change should be managed by professionals to yield better results and add enhanced value to the concerned organizations (Coleman 1990).

## Culture as Capital

Culture is what we use while civilization is what we are. If we are to understand conflicts, stress and negotiations in business in a somewhat new light then we also have to appreciate the role of culture in the politics of everyday life of organizations where individuals engage themselves in pursuing different forms of economic activities.

Culture as capital can contribute effectively to the body corporate by defining the various role performances of the multiple stakeholders concerned in industry. Both the management and the workers can experiment successfully with different types of cultural modes relevant to their industry.

Senior business leaders in India occasionally make it a point to organize brainstorming sessions where even junior managers, worker representatives and Trade Union leaders can voice their own ideas about better productivity, new R&D, innovative marketing, revolutionary designs and improved economy of scale.

So it appears that both choices and voices have to be entrenched in the domain of better business sense boosted by motivation, inspiration, trust and cooperation. The more broadbased this industrial dialogue happens to be, the better would be the chances to experiment in a more creative manner with the changes that lead to overall better management of business organizations.

## Managing Change

The management of change is a critical discourse. It takes place at the levels of perception, motivation, attitude and leadership. How best to motivate both the management and the workers to coordinate business activities under a different regime of change is often a fascinating study that can lead to exciting new insights related to the organic realities of organizations. How people try to cope with stress and strain, conquer their anxieties and worries, battle with their learnt values against new values and finally emerge as either winners or else losers is a crucial exercise that impinges upon the working of processes, products and arrangement of realities both within and outside organizations.

Change Management is basically a function of strategized cultures pitted against fractured realities. The ways in which the owners, managers and workers perceive reality / realities may be radically different from one another. Their perceptions are conditioned by their Culture Root Paradigms that constitute the overarching rhetoric of change of their organizations. Change can both create and destroy.

So Industrial Relations as a dynamic of Business Organizations can also serve as a metaphor of change against the overall backdrop of ethical applications in business. How objectively to wield any given technology, how to cohere and conjoin angularities in attitudes, how to add to the existing stock of Social Capital by sustaining working relationships in business etc can arguably constitute the first steps towards building what social scientists and field practitioners engaged in the Development Sector tend to address as Multi-Stakeholder Dialogues.

## Multi-Stakeholder Dialogues

Such a dialogic mode of interaction involving multiple actors (and their roles) as well a different stakeholders (with their own typical interest/s in the organization concerned) is not always easy to institutionalize as a new credo of organizational reality. Tripartite talks between the Government, Management and labor may be cited here as a rudimentary example of such multi-level, multi-faceted and multi-functional dialogues. But Labor Laws and social welfare legislation apart, these dialogues are essential to understand in consensus the aims, aspirations and both long- and short-term business strategies of organizations.

One of the basic objectives of multi-sectoral stakeholder dialogues is to move away from the merely welfarist, altruistic and pluralistic orientations of organizations to a given mode of multiculturalism that has to be learnt, lived with and assiduously practised in order to transcend everyday realities to an aura of a changed way of business life. Post-postmodern societies have to desperately try and learn the post-global manner of existence where almost nothing is permanent, so to say. Markets change and clientele change while entire economies go through phases of radicalization that tend to impart new meanings to business strategies and ethical applications.

So the entire gamut of signifiers and signifieds also change and cater to further transitions in business organizations stylized and dictated by the polyphony of voices. These voices are supported by choices of change, regime of discourse and strategies of organizational development. The plethora of voices is an indicator of change in terms of both process and progress, although these are rather value-loaded expressions. So the more vibrant dialogues become the more adequate is their scope to include. Because if modern business today happens to stress increasingly upon Sustainable Development and Inclusive Growth, then it also becomes a necessary commitment to emphasize more on the broadbased nature of quality and multiple dialogues that encompass the small narratives of power and authority.

## Ethical Business Applications

The combination of business sense with ethical sense is a utilitarian, altruistic as well profiteering strategy that is difficult to follow conscientiously in the real world. Individuals are not always driven by ethical aspirations. It is a given fact of life that we generally follow the predatory instincts of our personal profit motives. The challenge is to inculcate the best possible fit between profits and ethics, immediate gains and strategic alliances, present temptations and future goodwill. A modicum of balance has to be struck in this context so that the different types of extraneous and internal variables that tend to influence such a process of decision / policy-making can be balanced in an almost rational manner.

Our ethical selves may not always be our instinctive beings. It is a metaphysical question of existence versus sustenance, learnt values versus inherent proclivities and cost-benefit analyses between the self-satisfying self and the social satisfying other.

When such conflicts between the otherness of the other and the openness of the self tend to be reconciled in the context of specific industry and general society, then the other uncomfortable questions regarding globalized dichotomies, the Digital Divide and the inchoateness of the self versus the nascent sense of otherness emerge.

## Business Sense

So it would appear that good or bad business sense impinges more upon the multiplicity of dialogues rather than decisions taken in isolation. So the analyses with regard to processes and progress necessarily indicate whether business decisions and / or solutions would be sustainable in the long or short run. An examination of perspective is important to both assess and understand the criticality of business sense in the context of improved Industrial Relations.

So how does one ensure better relations in business while at the same moment trying to make higher profits? This decision can only be made after a Knowledge Attitude Perception study is undertaken to evaluate the comparative benefits / estimates between better business relations and higher profits. Ethical applications in business activities should also be actively considered in this connection.

## Sociability Discourse

So finally we are led to the direction of the sociability of business i.e. how social is business and how widespread and traumatic/therapeutic are the remedial effects of value applications to any given industry. Are workers now being consulted, are their priorities and preferences being listened to, are they more confident in the management, bolstered by the thought that retrenchment and lay-off would not be enforced until absolutely imperative? Power plays become operational at this juncture when not only the social and cultural aspects of business but also the political and economic facets of the industry have to be accounted for to study the art of resolution of conflicts at the workplace.

## Conflict Resolution

Conflicts of interest that arise from clash of values, attitudes, perceptions, orientations and values are occasionally constructs of different cultures. The manner in which actors engage in conflicts and lock horns with each other over issues of preferences and priorities often emerge from skewed realities where the concerned parties are located at different levels and on different levels of cultural ensemble and trappings of power.

Because conflicts primarily begin at the level of knowledge where the mind is constantly busy weighing different options. The manner in which individuals approach conflicts is also rather culture specific. The issue of Industrial Relations can be better understood if the fragmented concerns of conflict are better appreciated. Conflicts at the higher echelons of power are more about abstract/intangible realities such as power, authority, prestige, esteem, influence and access to the repository of high-end corporate values.

## Insights/Enigmas

The insights dwelled upon in the course of this paper and the enigmas studied tend to open up an entire new discourse of possibilities and politics that correlate Business Sense, Corporate Social Responsibility, Social Capital and Ethical Applications.

The more intricate and intrinsic happens to be the question of Industrial Relations, the more complicated becomes the entire debate of Social Capital. Because the specifics of trust, value and cooperation predicated by the realities of corporate networking and social embeddedness are inherently highlighted in the context of better performance, enhanced interactions and complex commitments of the industry to the market, clientele, quality, pricing and Corporate Social Responsibility.

## Conclusion

While concluding this paper it is important to retain the lessons that were learnt and the best practices identified during the theoretical discourse of drafting our original argument about

Business Sense vis-à-vis Social Capital and Rational Choice theory.

It lies in the rational interest of the concerned stakeholders to cooperate among themselves, make collective decisions and approach both the present and the future in a consolidated manner. This can be a viable way in which to sustain organizations in the face of the threats triggered by Globalization. But these threats can also be translated into potential opportunities.

The Market, Civil Society and State happen to interact in a dynamic manner that also tends to convert Weaknesses into Strengths in a post-globalized world. But the mode of inclusion of Social Capital theory within the ambit of the discourse of Industrial Relations is a new benchmark that may lead to interesting new research in this area and collate new theories in the light of new facts and new possibilities.

## Acknowledgments

The author wishes to thank his colleagues at the Danish Institute of Rural Research and Development (IFUL) under the University of Southern Denmark, Esbjerg where he was invited as a Guest Professor in 2007 for their insights into Social Capital that originally inspired the first draft of this paper. The usual disclaimer applies.

## REFERENCES

Bjørnskov, C. (2004), 'The economic use of social capital: the income relation', In M. Paldam, and G.T. Svendsen (eds.), *Trust, Social Capital and Economic Growth: An International Comparison, Cheltenham*, UK: Edward Elgar, Ch. 7.

Beugelsdijk, J.A. (2004) "Bridging and bonding social capital: Which type is good for growth?" In: Arts, W. Hagenaars, J. & Halman, L. (eds), The Cultural Diversity of European Unity. Findings, Explanations and Reflections from the European Values Study, Brill European Values Studies, EVS 6. Koninklijke Brill N.V, Leiden, pp. 147-185.

Boix, C. and D.N. Posner (1998), 'Social capital: explaining its origins and effects on government performance', British Journal of *Political Science*, 28 (4), 686–93.

Bourdieu, P. (1979) 'Le capital social' [Social capital], Actes de la recherche en sciences sociales, 31, 2-3.

Bourdieu, P. (1986), 'The forms of capital', in J.G. Richardson (ed.) Handbook of Theory and Research for the Sociology of Education, New York, Westport, CT and London: Greenwood Press, pp. 241–58.

Burt, R.S. (2000). "The Network Structure of Social Capital". *Research in Organisational Behaviour*, Vol 22, pp. 1-83.

Burt, R.S. (2007) *Brokerage and Closure: An Introduction to Social Capital*. Oxford University Press, Oxford. Forthcoming.

Chloupkova, J., G.L.H. Svendsen and G.T. Svendsen (2003), 'Building and destroying social capital: The case of cooperative movements in Denmark and Poland', *Agriculture and Human Values*, 20 (3), 241-252.

Coleman, J.S. (1984) "Introducing social structure into economic analysis". *The American Economic Review* 74 (2), 84-88.

Coleman, J.S. (1988a), 'Social capital in the creation of human capital', *American Journal of Sociology*, 94, 95–121.

Coleman, J.S. (1988b), 'The creation and destruction of social capital: implications for the law', *Journal of Law, Ethics & Public Policy*, 3, 375–404.

Coleman, J.S. (1990), Foundations of Social Theory, Cambridge, MA: Harvard University Press.Kolkata 700 064, WB.

Social Capital Gateway (edited by Fabio Sabatini)

**URL** <http://www.socialcapitalgateway.org/>

Prasenjit Maiti and Gunnar Lind Haase Svendsen *et al* (eds), *Social Capital*, Vols. 1 and 2 (New Delhi: Atlantic, 2008).

Prasenjit Maiti, *Development Ethics* (New Delhi: Concept, 2008).

# 15

# Is It Politics or Profits or Even Compassion? Unraveling the Motivations for Corporate Social Responsibility

*Dr. Prasenjit Maiti*

## ABSTRACT

This article makes a politico-sociological attempt to analyze the real (?) factors that drive the apparently philanthropic notion of Corporate Social Responsibility (CSR) or Corporate Citizenship towards policy-level implementation and praxes. Various reasons such as image-building exercises, confidence-building measures, power politics, bargaining strategies, negotiations and compromises, altruism, accountability towards the concept of Sustainable Development (if not Inclusive Growth!) and Good (Corporate) Governance are examined in this process. The paper tentatively concludes that CSR is rather situation-specific: it relates to profits, power plays, politico-legal equations, situations of market flux, responsible images, humane concerns and other such ethical dynamics in a politically correct manner that can be perhaps best described as being inchoate *i.e.* CSR is issue-driven and context-bound but not merely *ad hoc* simply because it can also envision the larger perspectives, ulterior objectives and grander narratives of the corporate *longue durée* version of history available in terms of organizational growth and hegemonistic / institutional entrenchments. So the power anatomy of CSR is as hierarchical as the stratified and complex matrix of social relationships and sociability profits that sustain and help evolve the ever-changing practices of Corporate Citizenship.

**Key Words:** Corporate Ethics / Citizenship, Ethical Business Applications, Responsible Business.

## Introduction

CSR or Corporate Citizenship (also variously known as Sustainable Responsible Business or even Corporate Social Performance) has steadily emerged as one of the defining parameters of the entire arrangement of business strategy since World War II. The concept of social welfare has been integrated with business management for purposes we shall have occasion to interrogate subsequently in this paper.

But what is interesting at this juncture is to appreciate how in fact these divergent trends have been cohered to generate new ideas and images in the market place of corporate politics that has increasingly found it difficult to ignore the polyphony of voices and choices underpinned by the social agora of change management initiatives (Toenjes 2002). The multiplicity of business choices has to now necessarily conform to the obtainable arrangement of community-level developmental choices in order to institutionalize the best practices of social entrepreneurship and corporate citizenship.

This development towards new possibilities has largely been made possible by the emergence of the Civil Society (following Antonio Gramsci) as the Third Actor that has moved away from the confines of both the State and the Market to create a dialogic space for itself that also caters to stakeholders (and not merely stockholders) interacting within and outside the Public Sphere (following Jürgen Habermas) on the issues of preferences, priorities and polemics that tend to affect their politics of everyday life (Lehman 2001).

What is rather important to understand here is the fact that the Civil Society has distanced itself away from the Market while citizens limited to political boundaries (or netizens without any such foreseeable or apparent boundaries) have also been socialized enough to agree to disagree (not really in consonance with Noam Chomsky's thesis of manufactured dissent).

More intellectual debates and critical discourses at the social level of politics have largely been a derivative of development studies. The classical studies on perverted development had led to the formulation of the metropolitan center versus satellite analyses that indicate that drainage of indigenous wealth and natural resources from the South to the North has resulted in a top-heavy model of so-called economic development (Frank 1967 and Preston 1996).

Subsequent decolonization and establishment of postcolonial democracies in the South had only confirmed Frank's observations. Such a pattern of exploitation based on mercantile/industrial/finance capitalism and culminating in cultural imperialism logically led to the process now fashionably known as globalization (following Immanuel Wallerstein and Eric John Ernest Hobsbawm).

But globalization has also been challenged by glocalization with a mantra championed by the World Bank that urges us to Think Globally but Act Locally.[1] Added to this imperative are the Millennium Development Goals put forward by the United Nations Development Program.

So the question remains: is CSR a proactive or a reactive development that impinges upon the realities of social markets? Citizens and consumers have equally learnt to say no and look elsewhere for viable alternatives while the rapid expansion of the Development Sector has also ensured that Community-Based Organizations, Civil Society Organizations, Voluntary Organizations and Non-Government Organizations have gradually become sensitized to the intricacies and specifics of various multi-stakeholder dialogues that happen to inform the very fabric of consumerist societies and commodity fetishism (following Karl Marx).

So the war is now principally focused on the mind and market rather than territory and matter because cultural imperialism as the new form of colonization has invaded the sanctity of the Self by creating the deprived Other. The sale of a product as innocuous as Coca-Cola globally has provoked critics to come up with the concept of Coca-Colonization that simply implies taking the world by storm while equipped with the American cultural baggage that in itself is both a social as well as a power statement (Amba-Rao 1993).

## Hypothesis/Methodology

We propose in this paper that CSR as a corporate strategy is fundamentally a dynamic policy mechanism that facilitates better business profits by projecting to broad-base social profits accruing to the vulnerable communities being disproportionately affected by the processes of mass-scale industrial production.

Adverse impacts of degraded environments, toxicity in the air/water/soil due to pollution, alienation of the self of the worker from the otherness of his / her produced outputs, dissociation of the creative psyche from the produced entities, disjunction between labor and wages *etc* are sought to be rationalized in an arbitrary and *ex post facto* manner.

The methodology deployed in this paper conjoins content analyses, theoretical constructs, historical/archival data mining and analyses of the given primary facts in the intellectual framework of meta-narrative and post-modern historiography. Case studies have been investigated and contemporary theoretical baggages/underpinnings critically gazed at in order to understand the neoinstitutional and hegemonistic trappings of corporate power that happen to inform CSR while not at all reorienting the critical and traumatic relationships between individuals / beneficiaries (?) and their institutions / benefactors (!).

## CSR Politics

The politics that more often than not happens to inform the politics of CSR is multi-dimensional in character and multi-layered in its impact. It is somewhat similar to the politics of aid when donor countries or funding organizations from the North extend financial support to the South in exchange of disguised Structural Adjustment Programs (Bennett 2002).

The ostensible reason for extending grants or aid (not loans) is humanitarian in its scope and approach but the real reason is somewhat different. So a soft drink company from the North may well sponsor a safe drinking water project in a village from the South and then start marketing its product after having established a pro-people/pro-poor image. This is nothing else but vicarious marketing that is cloaked as ethical and responsible business practices (Jackson 1993 and Moriarty 2005).

The debate whether multinational corporates should at all involve themselves in social expenditures is rather old. A certain school (following Milton Friedman) has steadfastly maintained that business should only be concerned with making profits. But modern business is smart enough to understand that economic profits increase with an expansion in social and cultural investments because popular trust and favorable perceptions are then associated with the company's products and promote sales in this process (Elkington 1994 and McClintock 1999). But this entire process has to be sustainable and decided on the basis of viable business options.

So the issue is nearly all about fusing the identity of products (brand equity) with the public image of the company. The concerned company would ensure a steady level of profits and more once this is achieved. The question is really very simple: how can a company integrate its business objectives with its altruistic image for the purpose of increasing profits? CSR lends itself to a multiplicity of explanations in this regard as would be evident in the course of this article.

This apparently difficult task is simplified once a business enterprise makes inroads into a market by institutionalizing its business ethos *vis-à-vis* the society and the multiple communities. The Tata Group of Companies in India is a potential example that readily comes to mind in this connection: the Tatas have always projected a sense of responsible and socially committed business in a country like India by investing considerably in the area of social infrastructure development.

A modern city like Tatanagar has emerged around the Tata Steel and Tata Motors factories while the Tata Institute of Social Sciences, Tata Institute of Fundamental Research, Tata Energy Research Institute (now known as The Energy and Resources Institute) and Tata Memorial Hospital are a few important institutions in India in the public domain that owe their inception and inspiration to the Tatas.

The general perception in the popular psyche is that any commodity and / or service that has the Tata brand associated with it in any which way would be of endurable value. Tata Consultancy Services and Tata Indicom are next generation companies of this group that has steadily evolved with the times by not wavering from its core business values. Now Ratan Tata has come up with his people's car (Tata Nano) that is affordable while at the same time is also profit making for Tata Motors. So the idea is to portray a philanthropic image by targeting niche markets that would still ensure corporate profits while expanding the bases of the company's social acceptance (Booth *et al.* 2000).

## Monitoring & Evaluation

Monitoring is the systematic collection of primary data to provide the corporate management and primary stakeholders with an indication of project progress. Evaluation is the methodical and objective assessment of an ongoing or completed corporate project, including its design, implementation and results.

Evaluation leads to more informed decisions, allowing those stakeholders / stockholders involved in the profit-making project to learn from experience and to be accountable to corporates and stakeholders alike on the basis of identified best practices. Both monitoring and evaluation should be used to demonstrate accountability and to understand the dynamics of the project.[2]

Willingness to pay underscores the anticipated revenue earnings of the corporate management in terms of user charges fees so far as operation and maintenance of civic / infrastructural assets are concerned. Better supervision of existing infrastructure indicates optimum utilization of manpower presently available ie streamlining human resources and avoiding incurrence of non-merit expenditures (free rider subsidies) in this process. Such benchmarks examine the value for money concern of good corporate governance as a central issue of sustainable development strategies informed by stakeholders' consultation / participation and mobilization / information. Benefit mapping of CSR projects in terms of new infrastructure created is also an exercise related to physical progress measurement of engineering interventions and designs at the ultimate level of implementation.

Monitoring and evaluation indicators may be both quantitative and qualitative in order to include process indicators, progress indicators and qualitative indicators. These indicators provide data inputs to monitor the project, evaluate progress of engineering works and guide the corporate managers to ensure and safeguard efficient use of valuable and limited resources.

Monitoring and evaluation indicators may also determine the magnitude of project impacts on the quality of life of beneficiaries before and after implementation of civil works. Such indicators can also examine the increased level of civil societal participation and awareness-generation in terms of CSR-related project interventions.

## Why CSR?

The basic discourse of CSR makes an attempt to conjoin the market with civil society by making business more responsible and the community more responsive towards the realities of profit-making ventures that have to compete with one another in the critical area of social dynamics as well. It is the question of allowing the disadvantaged and under-privileged adversely affected communities "speak out" so to say in the great carnival of Multinational Corporations and their profiteering discourses (Spivak 1988).

The concerns before CSR are in terms of challenges posed by poverty, malnutrition, unemployment, illiteracy, corruption, lack of accountability / transparency, imperfect implementation of projects and skewed delivery of benefits, socio-economic-cultural infrastructure and preconceived notions / mindsets (Edmunds 2003). The challenge is to decide upon a harmonious and presentable order of socially committed business.

Management strategies are put to test while transforming these threats into opportunities and weaknesses into strengths. It is also a question of putting the right foot forward by identifying the areas that require the first attention of the company in order to carry the community together along the direction of participatory decision-making and profit-making efforts.

The company has to be seen to deliver social goods and services. This visibility factor assumes supreme importance in the face of dogged resistance posed by activist and pressure groups that militate on grounds of sustainable environmentalism and inclusive growth. It is true that natural resources would be depleted and environmental pollution would take place whenever industrial production on a large scale commences in any given area. But the basic idea is to initiate a process of consultation with the communities concerned from the angle of benefit mapping exercises (Gray *et al.* 2006). The idea is to convert pedagogics to deschooled and decentralized decision-making.

## Benchmarking CSR

CSR can be ideally benchmarked by putting into place several benefit monitoring and evaluation strategies at different life cycles of the company's social interventionist exercises. The different stages and angles of CSR benchmarking can be understood by the following model:

**CSR Benchmarking Model**

| *Stage* | *Angle* | *Beneficiary* | *Method* | *Outreach* |
|---|---|---|---|---|
| Communitarian | Community | Primary | Participation | Local |
| The company fundamentally consults with the concerned communities | | | | |
| Civil Societal | Civil Society | Secondary | Consultation | Regional |
| The interactive dialogues engage larger audiences | | | | |
| Institutional | Institutions | Tertiary | Intervention | Global |

The smaller and larger audiences now share a common understanding

## Ideational CSR

The idea of CSR can be explored in its tradition of responsibility and ethos of social commitment. It is as if CSR tries to achieve an egalitarian social order as best as it can and thus comes into conflict with the basic goals of corporate entities. But the idea can also be found in the larger issue of moral ecosophy where deep ecology and philosophy come together to question the political and even human rightness of the morality of large-scale production systems.[3]

So even CSR as a concept can be categorized as Marxist, Green, Feminist, Pink, Black etc depending on the people practising it and the people for whom it is being practised.[4] Marxist / ecosocialist CSR would stress more upon an equitable Mode of Production that would entail a just (re)distribution of the surplus value being created by labor that is irretrievably alienated from the process, creative joy, forces, techniques and tools of production.

Green CSR would necessarily highlight the need for eco-friendly business practices that also take into account the overall depletion of the physical environment and try to make amends for the same by way of community forestry, joint forest management, eco-clubs, nature trails and similar ventures at the collective level by co-opting both the natural and the built environments as stakeholders of corporate growth and organizational development in any given context of production. Ecologically sensitized CSR practices would have to necessarily focus on the cutting edge of research in sustainable environmentalism in order to transmit the correct messages to the correct markets.

Feminist CSR essentially views the primary stakeholders as the alienated other that has been penetrated by the corporate stockholders / shareholders and fights for an engulfment of the same by establishing a regime of equal rights and responsibilities. The worldview in this case is based on the notion of exploitation and a tradition of lopsided power-sharing dynamics (Gilbert 1986).

Pink CSR assumes a somewhat diluted socialist approach while Black CSR again takes up cudgels on the grounds of ethnicity and claims that corporates owe much more responsibility towards racial minorities who have been exploited disproportionately since time immemorial when colonial capitalism and slave trade had their heydays.

## Pragmatic CSR

But CSR is really much more practical than it appears. It has to factor in strategies and negotiations of rapport-building with the community, civil society, multi-layered governments, activists and media. At all levels the message has to go through rather clearly and distinctively that the company has spent a lot of resources in designing and implementing its CSR mechanisms.

CSR has to be not only dialogic but also pedagogic in that others should perceive the company as having invested considerable intellectual and technical inputs into the operationalization of its community facilitation and liaison enterprises. The "good work" done has to be suitably packaged and presented before everybody and posted on the company's interactive website so that the right signals are transmitted to the right people at the right moment.

CSR requires hard-headed policy decisions that are ruled by reason and not passion. Compassion may influence the entire process but only in a strict rational sense. For social work also requires professionalism and experts to deliver the goods and services to the people if the company is at all interested in doing so and expresses sufficient goodwill in this direction.

It is basically an egalitarian social exercise by delegating specialized assignments to the technical experts who can take care of contingent situations and deliver the professional services that help to identify the Unique Selling Points of CSR (Harmon 2004). For CSR (following David Easton) has now become essentially an authoritative allocation of social values in order to reorient the balances of power and benefits in any given society / community. The myopic vision of business applications has to now outgrow itself and assume a new language in the context of different signifiers and radical signifieds that lend themselves to technically new interpretations.

## CSR and Images

The Bill and Melinda Gates Foundation, Sir Dorabji Tata Trust, Ratan Tata Trust, Azim Premji Foundation, Rockfeller Foundation *etc* are a few examples of CSR praxes in India and abroad.[5] As Nagavara Ramarao Narayana Murthy of Infosys Technologies had one expressed the opinion that beyond a point wealth has to be used meaningfully for social purposes, CSR primarily seeks to reinvest business profits into the mainstream of institutions and individuals called society to promote images and entrench images of responsible and committed business (Hill 2005 and Hill *et al* 2007).

But even image building exercises comply with the regime of priorities and preferences and notions of power superiorities. How to spend CSR funds so becomes a question of political equations and multi-faceted games played out in the uneven arena of globalized markets in the name of fostering multi-stakeholder dialogues (Hsieh 2004). The image building enterprises of CSR projects can be better understood in terms of the model formulated below:

**CSR IMAGE BUILDING MODEL**

| | | |
|---|---|---|
| Commitment toa Higher Moral Order of Business | Accountability to the Environment and Ecology, Sanctity of Human Life and Labor (both Physical and Intellectual), Responsibility towards the Systemic Mode of Production | Initiative on a Local Basis to Conform to Global Standards |
| | Change Management in order to Ensure Consonance with Social and Business Profits | |
| Innovation as Capital | Transparency and Sociability to Creatively Outgrow Images | Compulsions of Competition |
| | Good Corporate Governance[6] | |
| Socialization as a Resource and Sustenance | Participation and Dialogues and Pedagogics | Tradition of Responsible Business |

## Corporate Ethics

The issue of corporate values conjoined to the various narratives of corporate citizenship is more often than not predicated by the choices and priorities of infrastructure augmentation and enhancement of physical assets for public / welfarist purposes. The priorities of ethical choices applied to the sanctity of sustainable human development and inclusive growth emerges as a critical policy imperative in the larger discursive context of moral emancipation. This leads us to the larger and more problematic question of human rights when considered from the angle of civil societal liberties supported by CSR initiatives. How ethical can community development ultimately become when analyzed in the light of assumptions informed by corporate hegemony and the power / knowledge discourse that is inspired by the productive forces of the capitalist mode of production?

So sustainable development championed by CSR efforts often appears to be a panoptican construct (following Michel Foucault) that is cohered by a gaze that is transnational and transcommunal in its very scope, rationale and purpose.

Human development underpinned by corporate citizenship can never emerge as a value-free project that is not conditioned by the microphysics of capitalist power or even the archaeology of meta-knowledge. The semantic of corporate learning is an ontologic exercise that cannot be unduly coerced by the versatility of multicultural complexities or pluralist developmentalist politics.

We have to appreciate the fact that the overall dynamic of ethical business applications cannot really be undermined irrespective of whatever value connotations development may subsume at different stages of infrastructure augmentation projects that happen to impact different categories of stakeholders in variant magnitudes and qualities.

## Ethical Business Applications

Redefining stakeholder entitlements is a process that intrinsically challenges predominant corporate assumptions in the context of growing pressure of population and fragmentation of urban resources. New material conditions of life are reoriented and differently molded in this process. Lesser payoffs in terms of minimum tangible benefits to primary stakeholders may take place as a direct spin-off of the multipliers of poverty reduction programs. This is somewhat similar to the diminishing marginal utilities of consumption and over-utilization of resources that finally lead to consumer satiety.

Poverty alleviation initiatives and income generation measures may lead to gender empowerment for a certain section of the community. This may or may not lead to formation or facilitation or even enhancement of the available stock of social capital so far as the target groups among the community are concerned. New culture root paradigms may get entrenched in this critical process as functions of new societal arrangements and perceived specifics of realigned identities and different matrices of power. Such may be the case when new local markets are sought to be explored in the face of established global markets.

Such a state of affairs would require new institutional networks that may work in a milieu of change management predicated by sustainable development alternatives. This necessitates a new regime of discourse of dynamic CSR where the terms beget different signifiers and alternating signifieds. Voices and choices of the people would become confused otherwise in a cacophony generated by the multitude or the Tower of Babel of corporate policy alternatives. Such an overwhelming otherness would subsequently provoke dystopic and dysfunctional social pathologies that would be without any incipient tenets of hospitality (following Jacques Derrida) whatsoever.

Systemic compulsions of the new international economic order ushered in by the World Trade Organization and the General Agreement on Trade and Tariff may be viewed as in-built disruptions that only add background noise to disturb channels of communication between the social market and the political consumer who is prepared to utilize significant and limited resources in order to access utilities. Depleted capacities of primary stakeholders occur as a result of the systemic imperatives of globalization and CSR.

## Responsible Business

Populism is another disabling factor that plagues the ideology of CSR. We come across business policy decisions that are not commercially tenable in the market but are implemented nevertheless due to the compulsions of so-called pro-people and pro-poor politics that is not akin to dynamic CSR strategies. Local actors may be able to reorient a new brand of corporate politics in their favor by manipulating primary resources available at the grassroots. But we have to deal with local as well as macro-level corporate politics simultaneously in order to coordinate between the compulsions of subalternist and elitist power approaches.

The CSR approach to social realities may ultimately serve to restore the community at the center of its developmental focus but at a significant cost. This cost is the one incurred during informed exercises pertaining to social choice. So it appears that the incidence of poverty and the eradication of poverty are often to be studied in a context of change management.

Contemporary social realities would suggest that poverty becomes institutionalized as a phenomenon in the interest of powerful corporate actors who are able to obtain the maximum mileage out of the sense of relative deprivation that afflicts civil society. Poverty is also a function of social asymmetry.

This is due to the fact that social clusters are located along the economic hierarchy in different degrees of embeddedness. Public action and civic engagement in an enabling corporate environment may or may not entrench channels of cooperation among primary stakeholders. Such vertical divisions among society are caused by differing axes of participation and priorities conditioned by dynamic variables such as knowledge, attitude and perceptions.

We may increasingly appreciate that globalization is not only about processes but also about mechanisms that impact our daily lives and the manner in which we would prefer our lives to be impacted upon. This contention may appear to be somewhat irreverent and irrelevant but is merited with a certain quality of objectivity as well. So the nuances and criticalities of existence become all the more difficult to negotiate while rooted in one single destiny conditioned by one given and preordained order of corporate values. We require more values and non-values in order to co-habit with more meanings and more non-meanings. So it is merely all about conjoining abstract signifiers with concrete signifieds as the fashionable *glocal* way of life.

The contemporary processes of globalization do not only happen to constitute polemic engagements with *laissez-faire* utopianism because of the political specifics available in the First as well as in the Third Worlds but that most of the benefits of globalization also accrue to the First World because of its dominant policy to resist *laissez-faire* market arrangements in the movement of goods and services and labor forces around the world - especially into its domain - and its sustained opposition to the norms laid down by the Washington Consensus.[7]

Rich countries religiously maintain their immigration barriers and discourage agricultural imports while most poor countries have not been quite successful to attract much Foreign Direct Investment due to misgovernance on the part of their national governments. Rich countries, however, may as well concede that CSR politics is a fundamental informant / evidence of economic inequality since they are not likely to lower their agricultural and immigration barriers in the near future in order to facilitate protectionism at home. Rich countries may also review the performance of the Washington Consensus, which assumes that free markets necessarily promote economic convergence and underscores important issues such as the rule of law, property rights and transparent banking systems to sustain dialogues between the North and South.

The dynamics of CSR are more often than not perverted by the First World's studied reluctance to apply identical standards of *laissez-faire* to its borders and imports that it tries - with varying degrees of success - to impose on the Third World's legal structures and political arrangements. New global bodies such as the World Social Forum have also emerged in the recent years to meaningfully negotiate with the World Economic Forum and its politics of hegemony that extends to identities, cultures, languages, material conditions of life, ethnicity, social baggages, epistemology (knowledge) and ontology (existence).

## Corporate is Social

May we now shift our polemic stance somewhat to reflect that CSR-induced development may not always lead to freedom? Basic human rights (if guaranteed by CSR strategies) may lead to capacities on the part of citizens / netizens. But what about capacities that do not sell at the globalized marketplace? If globalization, on the one hand, facilitates decentralization then, on the other, it also helps develop pockets of dynamic Free Trade Areas in large developing countries such as China and India by reorganizing their economic geography, Foreign Direct Investments and global commodity supply chains.

This process, however, creates large hinterlands of economic backwardness and entrenches economic inequality *within* the developing South. CSR-induced development, therefore, intensifies regional disparities *within* the Third World. The so-called Internet Revolution is indicative of the hegemonistic battles waged on a daily basis around the world with reference to critical discourses of production, consumption, distribution and exchange. The World Wide Web was originally designed as a virtual space to permit efficacious and intuitive transfer of data across an array of proprietary networks but it is now generally deployed to intensify uneven power relationships.

## Multi-Sectoral Dialogues

CSR dialogues are expected to facilitate cross-sectoral stakeholder commitment and involvement in infrastructure development projects. Such dynamic interactions may also contribute effectively to streamline timely implementation of the physical works of development projects. Stakeholders' involvement in terms of public information and consultation exercises may extend to generation of awareness about the proposed physical works under different CSR-driven projects, active participation at the design stage and propagation of knowledge with regard to the timely implementation and cost-effective management of project activities.

## Social Capital

The popular understanding of Social Capital as an embedded civil societal resource supported by networks of trust and cooperation should be analytically conceptualized here in order to locate the altogether intriguing notion of corporate ethics in its proper intellectual discourse. Social Capital is more often than not underpinned by manufactured (or else fractured) consent that in itself is conditioned by parallel movements of power and culture identities.

What is trust and how can the intangible value of trust be defined by cooperation and multiple axes of social synergy that is cohered by an equally unquantifiable category called enlightened corporate self-interest that is variously known as faith?

The comparative value of trust when measured against infidelity may be quantified in order to understand why people sustain a working stock of Social Capital at the marketplace of CSR, in politics, public offices and the government or elsewhere where interactions and institutions come together to define various interfaces of the public domain.

To disown my neighbor in his / her hour of need or crisis may not be quite a wise policy when located in a broader temporal perspective. But the problem with rational consumers interacting at the level of the marketplace of CSR politics is that they are not always endowed with the gift of foresight that in itself is an asset conditioned occasionally by experience *i.e.* acquired social knowledge.

Why is the notion of Social Capital at all relevant in our community life informed by CSR initiatives? Is this because it is a useful and practical idea that is mutually beneficial in utilitarian terms? You would benefit if I am truthful while I would gain if you are honest. We can translate and extend this notion into areas of Corporate Good Governance such as accountability and transparency. The social structure as a whole has a developmental stake rooted in the intangible yet immensely valuable resource of Social Capital. This argument is substantiated in the case of corporate ethics.

## Conclusion

CSR-induced development, it may be appreciated at this point, is not anything extrinsic like politics imposed from the above without any regard whatsoever to the end-users of limited socio-economic and political resources. Actors who are supposed to interface with their very own institutions are nearly always better comfortable if left alone with the material conditions of daily life that breed organic ethos of community existence. This is where the colonial masters went wrong in Asia, Africa and South America when they bled the colonies white and left behind a legacy of comprador bourgeois and crony capitalism that, in turn, fostered a repressive state apparatus and a perverted anti-people bureaucratic managerial state system that was not only anti-people but was also occasionally anti-progress. So CSR politics is largely similar to the distressed Sphinx that appears enigmatic without any mystery and that is also rhetorical without any enabling strategic figures of political speech.

## NOTES

1. "Clearly, globalization, democratization and decentralization go hand-in-hand. As they do, they create new opportunities and new risks for all actors, including those who act at the subnational and local level. In effect, globalization and democratization create a greater incentive for effective policies and management than ever before and, as an inveterate optimist, I believe that the discipline they impose is a valuable instrument which will over the long haul ensure that national and subnational authorities behave more sensibly. The lessons we can draw from experience, as to what makes for an effective intergovernmental framework and for efficient subnational government are useful in that they can help reap the benefits from globalization more quickly and may help avoid the risks that globalization also poses. As we enter into the discussions of the next two days I would encourage us all to 'think global', but to focus on how we can more effectively 'act local'." (Quoted from the speech delivered by Johannes F Linn, Vice President, Europe and Central Asia Region, The World Bank, at the Global Conference on Capital Markets Development at the Subnational Level in New York City on 16 February 16 2000.)
2. See *Handbook on Monitoring and Evaluating for Results* (New York: UNDP, 2002) in this connection.
3. "Political ecology does not amount to a new program for intellectual deforestation, rather it is a historical outgrowth of the central questions asked by the social sciences about the relations between human society, viewed in its bio-cultural-political complexity, and a significantly humanized nature. It develops the common ground where various disciplines intersect." See James B Greenberg and Thomas K Park, "Political Ecology", *Journal of Political Ecology*, Vol.1 (1994).
4. "If ecosophy is to be a science, it is a value-laden one. If it is literature, it is an experimental genre that knows our language is presently not prepared to speak emphatically of our link to what surrounds us. If it is politics, it tends toward the utopian. Those of us in the midst of it do believe ideas can change the

world. And we are constantly reformulating, testing, expressing, looking for more roots and branches of these ideas in the past, the present, and the future of cultures all across the globe." See David Rothenberg (1994), "Beyond Back to Nature: Ecosophy Reinvents Itself as Always", *Trumpeter* 11(3).

5. "Health researchers have questioned the funding decisions and priorities of the Bill and Melinda Gates Foundation's multibillion-dollar effort to improve global health. The Gates Foundation, the world's largest philanthropic initiative set up by software czar Bill Gates and his wife Melinda, has committed nearly $10 billion (Rs 49,290 crore) over the past decade to organisations worldwide. It supports projects to make treatment for HIV, malaria, TB and other diseases available and to deliver vaccines against life-threatening infections. It also funds research aimed at introducing innovations in medicine. But two groups of researchers have said that the Gates Foundation is focusing on technology-based solutions, such as vaccines and drugs, and not investing enough to promote simpler, proven alternatives to tackle the health problems of the poor. The researchers have acknowledged the foundation's role in dramatically increasing resources available for global health. But, in papers published in the medical journal *Lancet*, they said that much of the foundation's grants go to western institutions and limited funds are available to build capacity for improving healthcare in developing countries. David McCoy and his colleagues from the University College, London, have cautioned that the foundation, through its influence on health policy networks and think tanks, may be indirectly undermining the decision-making capacity of health ministries in the developing countries." See GS Mudur, "Gap in Gates Giveaways - Health Researchers Question North-South Divide in Funding", *The Telegraph*, p. 1, 12 May 2009.

6. Good governance parameters include participation, consensus, accountability, transparency, responsiveness, efficacy, equitability, inclusiveness and the rule of law.

7. "The Washington Consensus as I originally formulated it was not written as a policy prescription for development: it was a list of policies that I claimed were widely held in Washington to be widely desirable in Latin America as of the date the list was compiled, namely the second half of 1989. Of course, development was the main objective of the countries in question; the point is that my agenda excluded policies even if I believed they would

promote development unless I was also convinced that they commanded a consensus." Quoted from a lecture ("The Washington Consensus as Policy Prescription for Development") delivered by John Williamson in the series "Practitioners of Development" at the World Bank on 13 January 2004.

## REFERENCES

Amba-Rao, S. C. (1993), "Multinational corporate social responsibility, ethics, interactions and Third-World governments: An agenda for the 1990s", *Journal of Business Ethics*, Vol. 12, No. 7, pp. 553-572.

Bennett, J. (2002), "Multinational corporations, social responsibility and conflict", *Journal of International Affairs*, Vol. 55, No. 2, pp. 393-410.

Booth, K., Dunne, T. & Cox, M. (2000), "How we might live? Global ethics in a new century", *Review of International Studies*, Vol. 26, No. 5, pp. 1-28.

Edmunds, B. (2003), "The integration of developing countries into international financial markets: Remarks from the perspective of an economics ethics", *Business Ethics Quarterly*, Vol. 13, No. 3, pp. 337-357.

Elkington, J. (1994), "Towards the sustainable corporation: Win-win-win business strategies for sustainable development", *California Management Review*, Vol. 36, No. 2, pp. 90-100.

Frank, A. G. (1967), *Capitalism and Underdevelopment in Latin America: Historical Studies of Chile and Brazil*, Monthly Review Press, New York.

Gilbert Jr., D. R. (1986), "Corporate strategy and ethics", *Journal of Business Ethics*, Vol. 5, No. 2, pp. 137-150.

Gray, R., Bebbington, J. & Collison, D. (2006), "NGOs, civil society and accountability: Making the people accountable to capital", *Accounting, Auditing and Accountability Journal*, Vol. 19, No. 3, pp. 319-348.

Harmon, J. (2004), "Introducing John Rawls: A review of *The Cambridge Companion to Rawls*", *Philosophy and Social Criticism*, Vol. 30, No. 5-6, pp. 643-663.

Hill R. P. (2005), "Do the poor deserve less than surfers? An essay on the special issue on vulnerable consumers", *Journal of Macromarketing*, Vol. 25, No. 2, pp. 215-218.

Hill, R. P., Felice, W. F. & Ainscough, T. (2007), "International human rights and consumer quality of life: An ethical perspective", *Journal of Macromarketing*, Vol. 27, No. 4, pp. 370-379.

Hsieh, N-H. (2004), "The obligations of transnational corporations: Rawlsian justice and the duty of assistance", *Business Ethics Quarterly*, Vol. 14, No. 4, pp. 643-661.

Jackson, K. T. (1993), "Global distributive justice and the corporate duty to aid", *Journal of Business Ethics*, Vol. 12, No. 7, pp. 547-551.

Lehman, G. (2001), "Reclaiming the public sphere: Problems and prospects for corporate social and environmental accounting", *Critical Perspectives on Accounting*, Vol. 12, No. 6, pp. 713-733.

McClintock, B. (1999), "The multinational corporation and social justice: Experiments in supranational governance", *Review of Social Economy*, Vol. 57, No. 4, pp. 507-522.

Moriarty, J. (2005), "On the relevance of political philosophy to business ethics", *Business Ethics Quarterly*, Vol. 15, No. 3, pp. 455-473.

Preston, P. W. (1996), *Development Theory: An Introduction*, Blackwell Publishing, Cambridge, MA.

Spivak, G. C. (1988), "Can the subaltern speak?", Nelson, C. & Grossberg, L. (Eds.), *Marxism and the Interpretation of Culture*, University of Illinois Press, Urbana, IL, pp. 271-313.

Toenjes, R. H. (2002), "Why be moral in business? A Rawlsian approach to moral motivation", *Business Ethics Quarterly*, Vol. 12, No. 1, pp. 57-72.